Portrait of an Island

SAPELO SOUND

SAPELO RIVER

BLACKBEARD ISLAND

DUCK POND

SAPELO ISLAND

ATLANTIC OCEAN

DOBOY SOUND

OLD TOWER

KEY

DRY LAND OR WATER

MARSH

TIDAL MUD FLATS

N

PORTRAIT
OF AN ISLAND

by Mildred Teal and John Teal

SKETCHES BY RICHARD RICE

Brown Thrasher Books

THE UNIVERSITY OF GEORGIA PRESS

Athens and London

Published in 1997 as a Brown Thrasher Book
by the University of Georgia Press, Athens, Georgia 30602
© 1964, 1981 by Mildred Teal and John Teal
All rights reserved
Set in Times Roman

The paper in this book meets the guidelines for permanence
and durability of the Committee on Production Guidelines
for Book Longevity of the Council on Library Resources.

Printed in the United States of America

01 00 99 98 97   P   5 4 3 2 1

Library of Congress Cataloging in Publication Data
Teal, Mildred.
Portrait of an island / by Mildred Teal and John Teal;
sketches by Richard Rice.
p. cm.
"Brown Thrasher books."
Includes index.
ISBN 0-8203-1961-9 (pbk.: alk. paper)
1. Natural history—Georgia—Sapelo Island.
2. Ecology—Georgia—Sapelo Island.
3. Sapelo Island (Ga.)—Description and travel.
I. Teal, John. II. Title.
QH105.G4T4   1997
508.758'737—dc21   97-19332

British Library Cataloging in Publication Data available

Portrait of an Island was first published by Atheneum in 1964.

To R. J. REYNOLDS, JR., *who made it possible for us to write this book, and to* VAL TEAL, *who encouraged us to do so.*

Contents

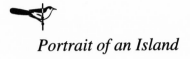

Portrait of an Island

Introduction

THE AIR OVER Georgia is dotted with spirals of gliding vultures, their wings outstretched as they cruise with power and freedom. No matter what their personal habits may be, they are birds of grandeur when flying. Be one of these vultures for a moment, a large, free-wheeling bird that can float hundreds of feet in the air on the currents rising along the beaches at the edge of the sea. You are wandering southward along the coast of Georgia as you rise on one thermal, then glide along to the next.

The May sun rose out of the sea two hours ago. Its penetrating rays warm the black surface of your wings, just as they warm the air to produce the updrafts which carry you along. At the top of an upward spiral, you can see the white thread of the coastal highway running beside a tidewater marsh, then cutting through a cypress swamp and a slash-pine forest. Occasionally the road bridges a river which drains the coastal plain, or spans one which has run all the way from the Appalachians.

On the seaward side of the highway you see the bright green, uneven belt of salt marshes varying from two to eight miles in width along the coast, separating a string of sea islands from the mainland. The marshes in areas which are cut by an intricate mosaic of small drainage creeks look bright green, while other areas are an unbroken, monotonous brown-green.

A short distance ahead you see one of the large sea islands, which forms the outer, visible edge of the continent. The beach rises gently from the At-lantic, preceded here and there by shoals over which the waves break and spray at high tide and upon which sea birds sun themselves at low tide.

On the seaward side of the island, at the north end, you see a large wedge-shaped section. This sec-tion is cut from the bulk of the island by a turning, twisting creek. Bands of trees running parallel along the tops of shallow ridges give the small wedge-shaped island a unique look, as though someone had trailed his fingers through the sand. A little to the south of the wedge-shaped island, on the large island, you see a long section of white sand beach behind which the trees are dying; the waves are eating into the land and almost through it to the marsh behind. To the south of this beach you see still another, running to the tip of the island; behind this one there are a very few trees, mostly small, and a few large, nearly bare sand dunes. The three beaches are separated from the main mass of island by salt marshes.

On the mainland side of the island is a large tidal river. Strung along it, in various sizes and shapes, are clumps of wooded islets or hammocks which are being eaten away by the tidal action of the river. You can see the swift current as the river carries rafts of marsh grass down to the sound, which empties the whole marsh system associated with the island into the open ocean.

As you fly from the north end of the island to the south, you see two large diked ponds; beyond these, large pine and oak forests, a little open space, and some tufted, thick grass savannah. In a few places, cypress trees grow out of standing water. Still progressing southward you see an airstrip to the west, large pasture lands in the middle, and a widespread settlement toward the beaches on the east side.

The whole land area is about twelve miles long and two to four miles wide. The main island is gently curled up like a withering leaf with raised areas along the edge and a depression in the middle.

Near the south end you see a square of large white stucco buildings surrounding a fountain spouting water from the lips of gnomes who are flanked by strutting concrete turkeys. Off to the beach side of the island you see another large dwelling set in a park-like grove of large live oak trees. At the very tip of the island is a red and white striped lighthouse, its paint aged, and its glass broken. It is abandoned.

Now stop, fold your wings, drop your vulture attitude. You have seen Sapelo Island and its wedge-shaped sister, Blackbeard.

This is the way we would like to have seen it on our first trip, not with the roar of airplane motors in our ears or by winding our way through the steaming, tortuous drainage creeks surrounded by shoulder-high marsh grass, but in a way that would let us see all the physical features of the island at once from a silent height.

We arrived at Sapelo Island in 1955, to work at the newly founded Marine Institute of the University of Georgia, and left it in 1959. We look back on it remembering its features, its changing disposition—stormy and sunny moods. We left with an appreciation for the struggles of countless animals and plants to live in an intricate system of interactions with each other, and for the importance of the surroundings to this system: the rocks and soils that make up the land, the weather, the moods of the sea. . . . We would like to give others a share in this appreciation—so we have written this book.

PART I

Sapelo

Sapelo Almanac

SAPELO LIES in a region with a mild, subtropical, humid climate. In the summer the sun is nearly overhead. Actually, it is overhead seven degrees to the south, over the Tropic of Cancer; but, nevertheless, the radiation is intense during a bright summer day. In the winter the sun moves 60 degrees to the south, at its farthest. It is then at an altitude comparable to that of the midsummer sun at the Pole.

There is little twilight compared to a more average American latitude, on the Great Plains or in New England, and the sun drops over the horizon in a matter of minutes around nine o'clock in the summer.

The temperature in summer ranges between 85°

and 95° F and in winter between 30° and 70° F. Winter days are rarely below freezing if there is no cloud cover. There may even be periods of balmy spring-like weather during the height of winter, which may suddenly give way to a more seasonable bluster.

During the summer there is a steady wind off the sea that brings both cool breezes and high humidity to the land. Under the intense July sun the temperature on the sand flats exposed by the tides reaches 120° F even though they are still damp with sea-water. The water itself reaches temperatures of over 80° F in the creeks and along the ocean beaches.

Along with the bright summer sun there is usually a stable high-pressure air system that develops over the Atlantic Ocean between the United States coast and Bermuda, known as the Bermuda High. This system pours a steady wind toward the coast which has the effect of preventing storms from moving into the area. There is a long period, beginning in April, when no storms visit Sapelo, that is, no large storm systems. Local thundershowers in some years pile up massive thunderheads every afternoon in June, pour rain, lightning and high winds onto a very limited area of marsh and island, and then disappear, usually by sunset. From a boat on the tidal river, surrounded by the flatness of water and marsh, one can see a storm descending on an area, pass by, and break out again in another spot. And, although there may be thunderstorms, the sun generally shines most of every day throughout the summer.

Off to the west, however, over the mainland, a squall line develops a few miles from the coast over the coastal plain. This gradually moves toward the coast as the summer progresses. When September arrives, this storm line reaches the coast; and in one week or two, Sapelo gets all of the rain it missed during the summer. We measured sixteen inches within two weeks one year. Since the island is only a few feet above sea level, it has little capacity to absorb rain, and the shallow, depressed interior of the island doesn't allow for run-off. The ground becomes saturated quickly. Water fills every depression, even those in the beach dunes. The island is awash.

About the time the September rains arrive, the hurricane season has started and the Bermuda High ceases to be important to Sapelo. The island is in danger, both from those hurricanes that come up the Atlantic and turn onto the land, usually between South Carolina and New York, and those that are born in the Gulf of Mexico and sweep across the Florida peninsula.

During the Plantation era and into the early twentieth century, severe hurricanes inundated the Sapelo —St. Simons area, completely ruining crops, taking many lives, and generally creating havoc. The lighthouse on the south tip of Sapelo was abandoned after it was cracked by a hurricane. Each year hurricanes threaten and usually veer off or pass by out at sea.

But each storm that passes sends out high tides, some destructive winds, and unsettled weather.

Generally, autumn weather is neither hot nor cold. Actually, one is not even sure when autumn arrives. There is very little leaf fall, since most of the island's vegetation is evergreen.

Winter announces itself with gray days of northeast winds, cold rains, and fogs over the approaches to Sapelo, which seem to make it infinitely more isolated than in other seasons when you can see the thin line of the mainland far in the distance. But spring starts early. A few warm days are always interspersed with the winter ones, and flowers begin to bloom in January. The spring periods lengthen, take in all of April, and summer begins in May.

Formation of a Sea Island

SAPELO ISLAND was in its place long before man stumbled onto it, but it is a new island as the age of the earth goes. With the whole lifetime of the earth in which to be born, some billions of years, Sapelo came into existence in the last or present epoch, the Pleistocene. It became established in its present form about the same time man came into being, a million years or so ago, and it is still growing, steadily pushing out into the ocean as sands are deposited along the beaches, and spreading back toward the mainland as the marshes silt in.

Let's begin several millions of years ago, before there was a Sapelo, when the region lay entirely underwater, as part of the ocean floor, viewed by fish

13

and turtles instead of vultures. Erosion, then as now, supplied sand to rivers, which carried it to the shallow coastal seas where it settled onto the bottom. As more and more sand accumulated, the lower portions were packed together by the weight above and cemented into soft sandstones by chemical action.

These sandstones never became as hard as granite or even as hard as the usual sandstones; a jackknife can be pushed into them easily, but they were strong enough to produce the sea islands. When these rocks and the sediments above them were raised above the sea level by shifts in the earth's crust, the sandstones were on the eastern edge of the new land with an area a few miles wide of softer sediments just behind them. As soon as the land was exposed, erosion began removing the soft sediments and eventually a shallow valley of salt marsh came into being.

The sandstones form the skeletons of the islands. The flesh that clothes this skeleton contains some remnant of the soft sediments, but for the most part it has been constructed by the sea since the island was raised. The building material consists of sands pulled up from the sea bottom by the waves and piled onto the outer beach. These small particles, once up on dry land, are available for redistribution by the winds which mold the final surface of the land before it is covered by plants and fixed in shape by their roots.

Waves break onto a beach in two general ways.

They may crash down almost vertically onto the sands, losing most of their power in the crash and washing up on the beach only a short distance before they rush back down carrying sand with them. These are destructive. Other waves come in at an angle, break gently and surge up the beach forcefully, carrying sand with them. They lose much of their water because their slower movement allows it to percolate down into the beach. The remaining water flows back down, gently retarded by friction, leaving behind the sand it carried up. These are constructive. Storm waves tend to be high and short, prevented from running up the beach by the backwash of the last wave, and are typically destructive. Good weather waves tend to be constructive.

In the case of Sapelo, constructive waves predominated in its history, and they were probably helped in their action by the changes in sea level that accompanied the advance and retreat of the continental glaciers during the Ice Age. The surf action moved out to the east when the water was locked up as ice, and the sand accumulated to the east was gradually transported up onto the island when later the melting ice released water into the oceans and the sea level rose again. Since the surf made several trips out and back during the Ice Age, more sand was piled up on Sapelo than would have been had there been a constant sea level.

The main body of Sapelo, the central portion with

the depressed center, wet and low, is the old skeleton. It is now piled high with sand dunes around the outer rim. On the west, the hammocks, Little Sapelo, Mary, Fishing, Pumpkin, and Jack, project from a large marsh about one third the size of the main island. This ends in the "creek known as Teakettle," as the original deed said, and Teakettle Creek still forms the western boundary of the island. The hammocks are outposts of the main island structure that have been separated from the main mass by erosion. The eroded depression is filled with marsh through which flows the Duplin River.

The Duplin is a special type, a tidal river, found only in areas such as the Georgia coast where the coastal lands are flat and where the tidal range is great. In a tidal river, which is small compared with regular rivers, there is almost no fresh water running from the land into the sea. All of its water comes from the rise and fall of the tides. As a result, the current runs upstream half of the time and downstream the other half, with nearly the same force coming and going. Whether it began with a straight or crooked course, as it aged, inequalities in the current strength tore at one side of the river more than at the other and the water meandered about nibbling away the hammocks on one side and the island on the other. The hammocks are appreciably smaller, some now less than half the size they were in 1760 when the first known map of the island was made by

De Brahm, a Colonial surveyor for King George III.

On the seaward side of Sapelo, the eastern side, sand has continued to accumulate, separated slightly from the main island by channels of deeper water. These accumulations of sand now form the three beach islets of Sapelo as well as the shoals offshore, while the separating channels have gradually filled in with salt marsh. The pattern of development is obvious on Blackbeard, where dunes of sand are still being added as the beach moves farther into the water. The new dunes are first bare, since salt left by the sea prevents the growth of land plants. After rain has washed the sand, conditions are right for the grass seeds which have blown onto the dunes to sprout and take root. The grasses are eventually followed by pine trees, which in turn are succeeded by oaks. All or most of Blackbeard was formed in this way, resulting in a series of alternating high, dry forested ridges and low, swampy or water-filled hollows.

The fast growing ridges of Blackbeard are being outdone by the faster growing south beach of Sapelo, but the growth features are quite different. From the early days of white man's invasion of the island, definitely since 1760, and probably since from fifty to one hundred years before, cattle have grazed the beach, preventing much growth of grasses on the dunes. And without these grasses, the sand isn't stable enough to build up into high dunes. As soon

as it is piled up by one wind, it is blown down again
by another. The lack of beach grass and the absence
of high dunes, which permit high tides to flood the
sand, also prevents trees and other plants from start-
ing growth. The commonest grass found on the new
dunes of Blackbeard was not found, or at least was
not common enough to be identified on the south
beach of Sapelo until the cattle were removed in
1959.

The middle beach, Cabretta Island, stretching
from the tidal river separating it from Blackbeard
to the inlet called the Big Hole, presents a different
picture. Rather than being built up, Cabretta is being
torn down, literally washed out with every tide. This
is not a matter of destructive waves but of lateral
currents along the shore which carry the sand lying
below low tide along the shore and dump much of
it toward the south where the waves carry it up onto
the fast growing south beach. As the sand below
Cabretta Beach is removed, the water washes the
beach sand down into its place and the dunes are
breaking down as the sea reclaims the sand. This is,
of course, the reverse of what must have happened
centuries past when the sea built Cabretta.

But now bark-stripped oaks and pines, bleached
silvery gray by the water are strewn, half buried in
the sand where they have been washed from their
beds. A thin, crescent-shaped ridge of tree-covered
sand is all that holds Cabretta to its beach form. If

the erosion continues as it has, very soon a break will occur in the ridge. Then the marsh behind and the beach will no longer be separated. What will happen in this event is anyone's guess. Certainly the soft marsh muds are much less resistant to erosion than the beach sand, and the area may wash out altogether leaving a bay.

As the sea was depositing sand on the Atlantic side of the islands, the rivers brought mud and silt which settled out in the quiet waters behind the island. When the mud flats that developed grew high enough above low tide so that *Spartina alterniflora,* the salt marsh grass, could grow, the salt marsh was born. The process can still be seen operating just off the south tip of Sapelo where a small island of marsh grass, only a few feet in diameter, is forming on the extensive mud flats exposed there at low water. This marsh island is seen at high water as a little tuft of grass separated very widely from any other marsh. Old timers on Sapelo remember when it wasn't there at all, and during the four years we were there, even a casual eye could see it grow.

This, then, is Sapelo Island, a central solid core that has been in its place for many centuries and is good for many more, only a few feet above sea level, bordered on the west by a great salt marsh and some small and disappearing islands, and bordered on the east by the Atlantic Ocean. The boundary between the two, water and land, is a wide expanse of white sand beach and a high ridge of dunes, which look stable enough from day to day but are subject to the whims of the seas and winds which rearrange the sands as they please.

The Invasion

As SOON AS Sapelo began to be, as soon as there was some land that stayed dry between tides, land animals and plants began to invade the island.

Although the salt marsh belt connects the sea islands to the mainland in a sense, there has never been a dry-land bridge. Plants and animals had to make their way across the barrier of marsh, tidal creeks, and rivers to colonize the island. For most species this presented no problem. The majority of plant seeds were small enough to be carried by the wind across the barrier or could resist a short immersion in water that might be brackish only during periods of high river flow. Some were carried by birds and deposited in their droppings. So land plants

21

arrived over a period of time on Sapelo.

Equally uncomplicated was the arrival of birds and flying insects, which merely winged their way over the short expanse. Spiders were carried by the winds as they hung from long threads of silk. Alligators, raccoons, and opossums made it handily by swimming through the rivers and streams. Somehow frogs and lizards made the crossing without, it seems, much difficulty, since all species present on the mainland are also present on the island. Perhaps they came floating on rafts of marsh grass stalks that had been washed out of their beds on the creek banks.

Strangely enough there are some species that have never arrived. Native mice found in dune areas and in back of beaches on the mainland are entirely missing. Also, the ground squirrels or gophers haven't negotiated the crossing. This is somehow without reason, because comparable small mammals did cross over. Moles are uncommonly common, and rice rats are plentiful. Wildcats inhabited Sapelo in the past, but either died out or were killed off by man during periods of intense cultivation.

Skunks, which do well on the mainland, apparently dislike wet feet so much that they never attempted a crossing. The blue-jay, which is more than abundant on the mainland, and even on St. Simons Island just next door south, is rarely seen on Sapelo. The jays' absence is probably due to lack of an urban habitat on Sapelo, or perhaps to laziness. The jay was willing to walk the causeway to St. Simons, but shirked the flight across the marshes.

In time most of the kinds of animals living on the adjacent mainland arrived. But there is more to colonizing an area than just getting there. The point is that animals have not only to arrive but also to find in their potential new home just the conditions that will satisfy their particular requirements as to habitat and food. And these requirements may be very particular indeed. Large animals tend to have rather broad requirements; and even if their preferred food and home should be quite specific, they can often move, if necessary, into another habitat and make do until their preferred home can be inhabited again. A typical forest creature may sometimes be found on the beach dunes or even in the swamps.

Small animals, however, cannot move far simply because their size will not permit them to do so. They can afford to develop specialized food requirements and achieve the efficiency that goes with specialization, since they require small amounts of energy to keep alive. But they must live in a place that fills their needs.

Extremes of these two types are illustrated by the tiny beetle that lives its life in an oak acorn, and men with their ability to live on a wide variety of foodstuffs, from acorns to elephants.

Sapelo has a great variety of habitats for an equally

great variety of animals. Forests of pine, forests of
oak, and forests of mixtures of these. Reforested
pine, old field pine on recently abandoned fields, impenetrable thickets of holly, raspberry, and horse-
briar, cypress swamps, freshwater and saltwater
ponds, open savannahs, marshes, beach, and even
such specialized habitats as old sawdust piles from
logging activities, where rough-winged swallows nest.

Complicating the food picture—who eats what
and where—are the seasonal and daily changes.
Some animals nest in one kind of vegetation and
eat in another. Some animals feed during the day
and others at night. So there is an ever-changing
population in the same spot, all overlapping in some
way. The day feeders aren't quite in their nests before the night feeders are stirring.

The populations of animals are dense and seasonal. At times there are population explosions when
thousands of dragon-flies swarm over the fresh-water
ponds and meadows; when a walk through the or-
chards stirs up clouds of small brown moths which
flutter like bits of paper ash in the wind; when sand
flies and mosquitos swarm out of the marshes by the
millions to make the few humans feel unimportant
and miserable, only to decrease sharply during a dry
spell and to explode again after a warm rain. At
times frogs, pollywogs, and chameleons are abundant; a few weeks later there seem to be only a few
adults in evidence.

We could not hope to count or discover all of the species on the island. We were there only four years; the discovery of every kind of animal might take several men twenty years of constant work to achieve. It would mean turning over practically every leaf and sifting through every inch of ground on the island. Still, in our four-year tenure, we discovered hundreds of species.

At least 225 kinds of birds visited the island during the four years. Many of these bred during the summer and left to winter farther south. Many stayed the year round. Some were only winter residents, but 225 kinds of birds makes an island busy—just with the bird population alone.

But there are also at least twelve species of frogs and toads, and these account for millions of individuals. There are insects; only the wildest guess could be made about the number of species. We saw and identified most of the species of mammals, many of the spiders and snakes, but an estimate of the vast numbers of beach creatures and creatures of the marsh could never be anything but speculation.

And to end this list of numbers of species, there is man, about three hundred strong, with a complicated pattern of relationship, division of labor, day-night habits, and food preferences.

PART II

The Land

Forest, Savannah
and Swamp

WE HAVE DIVIDED the habitats on Sapelo broadly
under three headings; the beach, the marsh, and the
land. The land is where the greatest variety of habitats occurs and where most of the species live. They
may wander during some seasons from the land
onto the salt marsh or beach, but for the most part
they prefer certain habitats and certain seasons. Our
"land" includes the forests, the savannahs, the
swamps, and even the fresh-water ditches and ponds.

The forests are of three main types: live oak, pine,
and cypress. Sapelo has great tracts of live oak forest,
each of which forms an almost continuous and
dense canopy overhead. They occur mostly on the
sandstone ridges which surround the central de-

29

pression. The trees are massive and beautiful. Some branches of the biggest ones may push themselves out forty feet from the trunk and nearly reach the ground with a few gnarled dips. The crown, the branches and leafy area of the tree, is usually much wider than it is high. The leaves, which are small, egg-shaped, smooth, tough, and shiny, grow in profusion, and they give the tree its name, live oak, because they remain evergreen. There is a general leaf fall in the spring; there is a sound like the sound of rain as the crisp leaves tick down and blow along the ground. The leaves that fall are the older ones. The young ones remain firmly in place. The leaf cover on the forest floor at this time may be two or three inches thick.

The ground in a live oak forest, especially in the one at the south end of the island where there is a long and continuous history of habitation, is clear and carpeted with a lush growth of grass. Under the shading canopy in other groves, a few species of weeds or an intrusion of palmetto or holly may achieve stunted growth.

The south-end oak forest is unique in that there are only a few old trees. Most trees are young and still growing in clumps of six trunks or more. This probably indicates that they grew as sprouts from a long-since rotted stump. The parent tree may have been harvested in the old shipbuilding days for frigates such as *Old Ironsides* (which was constructed from oak gathered from the sea islands of Georgia). Live oak was especially prized for ships because it is iron strong, even resisting marine borers, the most frequent attackers of hulls. Also, the lower branches and the joints of the tree, where the great branches are connected to the trunk, could be shaped into very strong, continuous ribs.

There is an old saying that a live oak takes a hundred years to grow, lives for a hundred years, and then takes another hundred years to die. Unlike the natural process of aging, a live oak looks old almost from the moment it begins to grow. It reaches a gnarled and powerful size early in life. A whole community of plants and animals take up residence along its branches. An oak tree is an island unto itself, providing both soil and water. The bark is very rough and deeply grooved. Along the lower, big branches, the bark catches and retains moisture. With time, a little dust and leaf mold accumulate in the wet grooves, soon making a fine soil, a kind of mulch, which is fertile territory for seeds or plants that are blown onto the branches. Even small trees, especially oak and palmetto, grow in the forks of the live oak trees. Resurrection fern takes to this dust and leaf mold especially well, and the ferns line the lower branches of most trees. The fern is green and erect during the humid, warm summers. It shrivels to gray curls during the cold, relatively dry winters, and expands again with the arrival of spring.

The trunk of the oak may be splotched with large patches of pink, red, green, and white lichens, and this increases the aged look. Strangler fig, grape, and wisteria vines embrace the trunks of many. Some of these vines have trunks themselves, four to six inches in diameter, growing parallel to the trunks of the oaks.

In the spring the grove at the south end of the island takes on a smoky lavender hue from the profusion of cone-shaped bunches of wisteria blossoms. A large, brightly-colored, orange bracket mushroom grows during the moist seasons on oak branch stubs or lightning scars.

The oak would look old by itself even without the addition of Spanish moss, or "Old Man's Beard," but the moss ages the tree until it looks like Rip Van Winkle after his long sleep. The moss hangs in great festoons, which may be so long and thick as to mute sound and still the air in the forests. It has a wonderfully clear, salty smell that clings to your hands for hours after you have handled the plant. The moss is gray-green and furry looking until a rain comes, and then, almost magically, it changes to fresh, light green as the fine hairs of the moss are plastered to the stem by the water.

Spanish moss is not a parasite, as is commonly thought; it only borrows a tree as a place on which to grow. It is somewhat choosy about its host and prefers oak trees, but it may also be found on living pines, sparsely, and on dead ones, profusely. It may hasten the death of a tree somewhat by covering the branches so lavishly that the host tree has difficulty pushing its own leaves through the dense mat of moss, but it does not kill by extracting food from the tree. The moss gathers moisture and minerals from air and dust, and from these manufactures food with the aid of chlorophyll and sunlight.

The pine forests on Sapelo are rarely clear stands, except those that have been reforested, and these stand row on row, inches deep in fallen needles. Mostly, the pine forests of longleaf, slash, and lob-lolly are interspersed with holly, oaks, vines, some cedar, and a lot of unidentified (by us), small trees. Much of the high ground was cleared for crops during the Plantation era; and after the collapse of this system with the end of slavery, the fields were abandoned. Slash pine moved into the cleared land, which was formerly in oak, and edged out the young oak struggling to re-establish itself.

After the first four years of putting most of its energy into producing a long tap root, slash and longleaf pine grows very rapidly. A tree may grow from sixty to eighty feet high with a clear, branchless, knotless trunk in thirty-five years.

These southern pines are as majestic as the live oak in quite a different way. They tower over every other kind of vegetation and have a relatively small area of greenery on top for feeding such a hulking mass of tree. The longleaf needles may grow up to eighteen inches long and give the tree a soft bushy look. The yellow flowers stand up like Christmas candles on the branches. All in all, it is hard to find a more beautiful tree silhouetted against a deep violet-blue evening sky.

Pine trees grow well only in fairly strong light, so they can't get started in the deep shade under an oak stand. They start easily in grassy fields, however, and grow in old fields before the slower growing hardwoods get established. Later, even when planted and well planned, pine stands may cause a dense

Clearing in the forest

shade, thus inhibiting growth of young trees. Most of the older natural pine forests are quite free of pine underbrush. Such conditions are exactly what young hardwoods and palmetto like, and so these get a good start and take over as the old pines gradually die or are cut. That is, they would take over but for one thing: fire. All through the southeastern coastal plain, there are dry spells in summer when lightning easily starts fires.

These fires, spreading through the woods, kill the young hardwoods; but the pine, especially the long-leaf, has a thick, fire-resistant bark and can live through all but the hottest fires. The lower branches

of the pines are shed naturally with growth, resulting in long clear trunks. These reduce the possibility of fire ladders occurring, which carry flames to the crowns of trees. Moreover, the young pines, struggling for a start in the forest undergrowth, also have resistant bark and buds so hardy they are known locally as "asbestos buds." As long as fires occur frequently enough, keeping down the accumulation of combustible material on the ground that might cause a very hot fire, the pines will thrive.

Palmetto also lives through relatively cool fires, which burn all of the fan-like fronds but leave the roots healthy to send up new fronds the following year. Consequently, the natural pine woods often have considerable palmetto on the floor. Repeated fires, which may leave healthy pine stands, will eventually kill the palmetto; but even with this check to growth, palmetto remains one of the most abundant plants on the island.

Before man came with his civilized notions of soil drainage, large stands of cypress grew in the interior swamps, pushing their knobby knees above the water for air. Then over a period of one hundred years, in a project started by one settler and carried on by succeeding ones, a drainage ditch, running nearly the length of the island and draining most of the swamp, killed almost all of the cypress.

After the plantations were abandoned, the drain-

age ditch gradually filled in so that water accumulated again in the low lying areas. Cypress, young
and still quite small, have once again begun to grow
between the carcasses of the still rotting old trees.
But the young trees have not yet formed the characteristic dense dark swamp that one may see on
the river flats of the coastal plain.

The drainage programs also destroyed the original
swampy grasslands, and their place was taken by
pastures cleared from the woods and by cattail
marshes made by diking salt marshes. The pastures
are dryer than the original grasslands, the cattail
marshes wetter. So each provides homes for some
of the original savannah inhabitants, depending on
whether the wetter or the dryer home suits the individual best.

Some areas in the cattail marshes are quite deep
and remain open water, and some are quite shallow
and completely choked by plants. Other scattered
ponds, mostly dry by late summer, and the drainage
ditches that still exist make up the remaining fresh
water areas. These form important living places for
visiting ducks, and for resident fish and alligators,
and as drinking places for other animals.

Finally, because the island has been inhabited for
centuries, there are habitats introduced by humans,
planned and accidental, current and from times past.

Larger Animals of
the Forest

ALMOST ALL of the world's available habitats are occupied by some species of animal. Some habitats support large numbers of a few species; others support a few individuals of a large number of species. On Sapelo, the salt marshes are made difficult dwelling places by the tides and have few species of animals, while the warm, humid forests provide quarters for a variety of forms.

The most unique habitat the area provides is probably Spanish moss. At least three species of birds, the parula and yellow-throated warblers and the painted buntings, nest in well concealed pockets in festoons of the moss. The painted bunting is one of the characteristic summer residents in the oak

38

woodland. Reference books invariably say this is a shy bird, difficult to observe. The Sapelo buntings are conspicuous, noisy, and even quarrelsome. The male's gaudy colors, a red body with a yellowish to olive-green back, topped by a purple head, make him especially obvious, flying up from the roadways, flitting through trees, sitting in the yards of the dwellings, or shouting at his fellows as they stake out their territories.

The female, a rather retiring olive-green-and-yellow bird, is much more difficult to see; and from the din created by the males, one gets the impression that the sexes are very unequal in numbers.

It is a general rule that during the spring migrations of birds to their northern nesting grounds, the local southern birds arrive first and the ones going farther on arrive later. This works out well; if all birds went north at once, those going farther north and only stopping over on Sapelo might find their nesting sites still under snow when they pushed on. The painted bunting is an exception. He does not arrive as early as he should, but turns up only after the northern warblers, thrushes, and sparrows have passed through. Then the buntings arrive in a mass. Perhaps, since they have time to raise two broods anyway, there is no pressing need to arrive earlier, and so they remain in the warmer sun, farther south. They come just in time to set up their territories, raise their young, and quite suddenly, around the first of September, launch out for the tropics again.

Painted buntings may be most characteristic of Sapelo oak woodlands, but many other species are present. Brown thrashers whir their way through the forest. Carolina wrens call cheerfully. The warblers and vireos flit around in the branches, catching insects; and an occasional, large, pileated woodpecker, with his bright red crest, makes a colorful splash, progressing from tree to tree calling with a voice proportional to his size.

During the day, in the dense, swampy woods near the south-end dwellings, one can often surprise a few chachalaca, or Mexican pheasant. This game bird, introduced on Sapelo by a former owner, sets up a really terrific racket during mating season. One male chachalaca will sit in a tree and call a few hundred yards to another who will answer in kind with a high-pitched, loud, exceedingly grating cackle. But even though his call might not make him nature's most desired bird, especially early in the morning, he is interesting to watch. In size he is midway between a pigeon and a ring-necked pheasant. The chachalaca

Chachalaca, or Mexican pheasant

often walk around the forest floor until they are disturbed. Then they take to the trees and sit, inexpertly teetering on a limb. The Sapelo flock has not increased appreciably, but it has maintained itself for about twenty-five years and seems to have taken to the new environment.

At night the calls of great horned owls carry from distant points in the woods. The sound seems to come from many individuals, but it is usually made by only one or two birds making their way from tree to tree, looking for night feeders or roosting birds. Whenever we heard the series of muted but powerful *hoos* of the great horned owl, we called in our cat, who was perfectly adequate food for the owls, and went out ourselves with flashlights to follow the bird around the oak groves. If the bird stopped hooing, we could always induce him to start again by imitating his call. On one of our stalking evenings, we saw an owl try to take a smaller bird, probably a dove, which was roosting in an oak tree. The screaming of the dove was pitiable, and the flappings of both birds frantic. The small bird escaped. The owl flew to a high crotch in a big oak and glared down at us. It could be that the great horned owl, like many predators, is rather an inefficient hunter in spite of his silent flight, large night-adapted eyes, and powerful talons; and that only weak, injured, or careless animals are caught.

Screech owls often take up residence in hol-

lowed-out limb sockets in the oaks or in abandoned
woodpecker holes. The ones we have seen on the
south end of the island prefer to feed on the paved
roads where large insects, rodents, and the like are
conspicuous against the unbroken asphalt background.

We found a screech owl nest in a rotted stump,
pitted with holes and undoubtedly overrun with
termites. It was a highly unstable home, but cheap,
with all the labor of pecking it out expended by the
former tenant, a woodpecker. There were three baby
owls inside, stacked one on top the other. The
mother was on top of the three babies, looking out
of the hole in typical owl fashion, glaring and un-
blinking. The top sibling was by far the healthiest.
The bottom bird, product of the late egg, was nearly
dead from starvation. The unobservant mother owl
never seems to know who has been fed and who
hasn't. She plays it by ear, and the baby with the
biggest mouth and most vigorous cry gets the most
food. The bottom bird, hatched from the late egg,
gets almost nothing. Actually, the mother can feed
only a limited number of children. The late hatching
egg is nature's extra insurance, only needed if the
early eggs are infertile; and so the baby birds hatched
from them are eaten or die of starvation if all the
eggs hatch—unless food is so abundant that the
mother catches more than the older siblings can eat.

In this nest, the poor little late-hatched chick

would soon die and be eaten or unceremoniously kicked from the nest. On his head was a large, old, many times reopened wound from the repeated peckings of his two bigger brothers.

We rescued this miserable child and raised him for a time; and in this case, he survived longer than did his brothers. We visited the nest days after we removed the chick and found it deserted. The babies were too young to fly, so they must have been eaten by a prowler, while our little screech owl, named Sir Toby Belch due to his peculiar gastronomic habits, made a dramatic recovery from his bad case of malnutrition and lived.

Like the great horned owl, the chuck-will's-widow is associated with the night in the oak woods. He is resident only during the summer, and his arrival is not noticed until he announces his presence with his plaintive call. Within a few days, so many of his fellows have arrived that ten different birds may be heard calling from different and often distant points. The territories are large, but so are the chuck-will's-widows and so are their voices. The initial *chuck* is soft but the *will's-widow* has a carrying power that makes it float through the air for a quarter of a mile or more. These birds fly through the night catching moths and other insects in their mouths, which form a funnel that literally stretches from ear to ear and is extended even further by a net of stiff bristles projecting from the corners. When day comes, the birds rest on logs or leaf-matted forest floors.

Wild turkeys now live in the mixed-forest areas. When the island was largely cleared for planting, the turkeys died out; but they were reintroduced and now thrive well, but with increases and decreases in the flocks. During two successive wet seasons we did not see a single poult, but the following year, which was favorable, warm and dry, we saw twenty, six of which grew to a size large enough to be incorporated into the flocks.

Turkeys are big birds: even in the wild they grow to be twenty-pound birds; and the impression they give is one of power. The commercial turkey seems a rather stupid, grayish bird; and although actually there is little difference between the wild and domesticated turkey, it seems natural to impute many grand human emotions to this majestic bird in the wild that are denied to him in the barnyard. The wild turkey has a rich bronze color, highlighted with greens and oranges. When he walks, he treads proudly. When he feeds, he feeds like a sophisticated gourmet. When he fights, he fights ominous, vital duels.

Wild Turkey

During the winter, the turkeys gather into flocks and forage together, there being safety in numbers. There were usually four flocks on the island. One was composed of about ten toms, another of thirty to forty females and three toms, and the two smaller flocks of one or two toms and less than a dozen females each. These four flocks had divided up the island between them, the smaller ones living toward the south end where disturbances from people are more common. As spring approaches, the males get intolerant of each other and begin to strut for the females and to fight among themselves. They have been feeding throughout the winter and have accumulated a heavy deposit of fat over the breast muscle. During the mating season, they exist on this food store. When the females are ready to nest, they disappear into the dense woods and build carefully hidden nests for their eggs.

Because there are no bobcats or foxes on Sapelo, the adult turkeys have no predators except man and disease; but during the egg and poult stage, turkeys are susceptible to snakes, hawks, and owls. Also, they catch cold easily if they get wet and are unable to dry themselves in the sun within a short time. As the poults grow, the flocks again begin to gather until the winter groups have assembled.

Deer, common in the island woods, have been there a long time, with little immigration from the mainland. It is the opinion of some mammalogists that the population has developed into a special sub-species, smaller than the usual white-tail. Perhaps something is lacking in the diet of the island deer that leads to the reduction in size. A full grown buck from Sapelo is only as large as a yearling from Minnesota. The deer on Sapelo, until recently, suffered heavy competition from cattle. Now that the cattle have been removed, the deer have been steadily making their way down from the wild north end toward the good grazing land at the south end, formerly occupied by the cows.

Hundreds of opossums waddle their way through the woods looking for anything edible: eggs, fruit, vegetation. A favored spot is on the roadways, where the animals are often seen and indicate that they are apparently too stupid to get out of the way of a moving object. Often we were almost thrown through a car windshield as we braked to avoid hitting an opossum calmly walking down the roadway ahead of us. The babies may be seen waddling after or riding on the back of the female, which is much larger than the male. About twenty minute young are born soon after mating. They are absolutely helpless and are transferred by the mother to her pouch, where they hang onto a teat, sucking and growing. In about five weeks they are large enough to venture out and ride about on their mother's back, but they nurse for a few weeks more before becoming independent.

The animal commonly thought of as a companion to the opossum is the raccoon. In general, both animals have the same habits. Both live in the woods, although raccoons often migrate into the marshes for tasty fiddler crabs. Both make dens in hollow logs. Both climb trees for protection and food. But in many ways, the two animals differ greatly. The raccoon is superior in every way to the opossum. He is much more handsome. His fur is thick and beautifully marked, and although the opossum is hunted for his scraggly pelt too, the incentive to hunt him is not as great. The raccoon is clever and intelligent, the opossum dull and stupid. The raccoon even tastes better.

All of these superior qualities have not paid off for the raccoon. The opossum steadily and hurriedly turn out their large broods, their numbers ever increasing as they blunder around, their very stupidity bringing them protection. The raccoons, producing only four to six young a year, are decreasing, for their superior qualities make them the prey of man.

We had always heard that raccoons made good pets if caught young enough. We were given one; and although time was against us, since he was already an adolescent with adolescent problems, we decided to make a pet of him. We named the animal Boris; we never really determined the sex, but Boris seemed an apt choice. He growled at us whenever we approached his cage. As Boris grew, his problems grew.

An agile climber, Boris scaled the broom handle, the refrigerator (from which he fell every day), the draperies, the inside screens, and even, seemingly, the bare walls. He spent most of his time somewhere near the ceiling, too frightened and unskilled to come down. He mewed pathetically like a kitten, but he refused help. We wore protective gloves when we dislodged him and invariably he attached himself to a finger. Discipline was no problem. No raccoon's-land for him was the table. One scolding accompanied by a small spanking taught him to stay off, at least while we were in the room. But Boris was destructive. During his residency, he broke a carafe, three antique plates, and a decanter. And like a puppy, he chewed up innumerable toys and papers and left dirty little raccoon foot prints all over the house. He was entertaining, but we felt only relief when he was led out to the back yard and took to the trees. For a time, he appeared regularly at the back door to be fed. One day we changed the brand of dog food we had been serving him, and Boris disappeared. Apparently the new brand was not to his taste.

Boris, the Raccoon

These are the more conspicuous animals of the forests, the birds and the mammals, large enough to be seen easily, active enough to attract attention. Some, especially the mammals, are secretive; but one can discover them, perhaps with lights at night or by following their voices. Furthermore, they have a kinship with men. They are warmblooded and intelligent. They are the animals in which men have assumed that emotions exist. However, they are vastly outnumbered and even outweighed by the thousands of species of smaller creatures and individuals of those species that live in these same forests.

Creeping and Crawling
Things

EVERYWHERE, on the land and in the water, are the creeping and crawling things. Some of the creatures spill over from the land into the water for breeding and crawl out again to live. Some take to the trees where they nestle in the moisture in the bark grooves. Others live under the ground-plant cover, or under moldering leaves. Some set snares for food with webs. Some rely on their skill as sprinters to get a meal. Others squat in one place, patiently watching with huge blinking eyes for delectable morsels to pass close enough to be snagged with deft tongues.

But everywhere underfoot is a great population of tiny, hidden, hard-to-see animals. One would have to spend years searching on his hands and knees before he could discover all the existing kinds of moths,

50

beetles, spiders, mites, and ants that occupy every kind of habitat: the moths which live as larvae only in the flowers of yucca; the dung beetles, which busily roll their medicine balls to a suitable location where they lay their eggs on the surface of the dung to provide an immediate supply of food for their young at hatching; the fishing spider, which goes onto the surface of the ponds in search of food that includes tadpoles and small fish.

The largest of the creeping and crawling animals are easy to see, but secretive. Once, crawling through a palmetto thicket, a friend came face to face with a huge diamond-back rattlesnake. This snake, which has little to fear, is not aggressive. He prefers not to take the initiative if he can move out of danger's way. But unaware of the snake's characteristics, our friend preferred to take the initiative. He left by the quickest route.

The rattler's venom is not the most deadly in the world. He has a blood-attacking venom, rather than the more deadly nerve toxin of the cobra or coral snake. However, because of the great size attained by the Georgia snake, he can inject so much venom with his protruding, moveable fangs that he holds one of the world's foremost positions as potential killer.

Since there are relatively few mammals on Sapelo, the food source for the rattlers is something of a puzzle.

The snakes that we killed and dissected appeared either to have been eating rabbits or to have gone hungry. Marsh rabbits are fairly abundant, and a number of snakes could make their living from them; but this leaves the food of the snakes living in drier areas unaccounted for. Opossums and raccoons are abundant, but it takes a rather big snake to swallow so large an animal whole. Cotton and rice rats occur in moist locations, the latter especially in the salt marshes. Moles are common, but other insectivores, shrews, and mice seem to be at most, very rare. We surmised that frogs and lizards supply an important part of the young snakes' diet, but we were never eager enough to know to study the problem firsthand.

Garter snakes, rough green snakes, and the colorful scarlet-king snakes, which mimic the deadly coral snake, occur but are rarely seen without specific search. Slender black racers were quite often run over as they tried to cross the paved roads.

We saw water snakes more often than any other snake. But only on a few occasions did cottonmouths or water moccasins appear around the duck ponds at the north end, their relative scarcity being attributed to the abundance of alligators there. The harmless water snakes draped themselves over the branches in all the small ponds, sunning themselves during the day. At night they would stare, gleaming-eyed, from the water when the beam of a flashlight was turned on them. Some were quite handsome, especially the copper-belly, whose burnished undersides contrasted nicely with his black topside.

The corn snake deserves special mention for three reasons: first, he is very handsome, marked with bold red splotches outlined in black on a gray or tan background, with the markings stretching out into dashing stripes on his head; second, the species tames easily and makes a good pet if properly cared for; finally, he feeds usually on rats and mice, an admirable diet for a snake, from the human point of view.

Snakes are common, but being generally secretive are not as obvious as their relatives, the lizards. Snakes hide from or threaten their way out of danger, but lizards outrun it. Being able to outrun trouble, they're not shy about exposing themselves in daylight.

The green anole, or new world chameleon, is the most abundant lizard on the island. On summer days, all sizes, from one inch to six, dart through the ivy on the island's buildings. One might lean against an oak tree and find six chameleons scooting out of the way. They dash along the roadways. Since they are relatively fearless, they can be more easily approached than other species of lizard. By combining

Corn Snake

a careful approach with a swift and accurate lunge, one can even catch them by hand.

The chameleon's ability to change color is not so outstanding or dramatic as that of a squid, but still, the process is interesting to watch. Chameleons change from bright green through brown to gray and almost white. The color is changed not to match the background but rather to suit the chameleon's mood or emotion. We kept a number of chameleons in a cage for a time. Their general color, when quiet, was brown with perhaps a bit of dull green. When an insect sweep-net load of assorted edibles was shaken out into the cage, things livened and brightened up considerably. The lizards became bright green and stalked their prey carefully. They captured insects, even flies, unerringly with a swift lunge and snap. Often, just before the final lunge, they would turn almost white with a few brown spots. Their food capacity was prodigious. A chameleon would stuff even a large grasshopper in until its body was distorted and the insect's shape was visible in the lizard's stomach.

Male anoles put on a special color exhibit in early summer. They become much more concerned at this time about trespass into their territories which consist of variable sized areas of ground, walls, trees, etc., in which they live and which they defend against other anoles. The color display, put on to warn other males or to attract females (the same display serves both purposes), is easy to observe during the mating season. When another anole approaches, the male does a few bobbing pushups with his front legs and the forepart of his body; then holding his head up, he expands a large rosy-red dewlap under his chin. If the intruder is a male, there may be a fight, but more usually the interloper flees. There is much seesawing in the boundary region between territories. A male is fearless and bold at home; not so belligerent, but firm, at the line; a coward when afield. The result may be that one male will chase another only to go too far, turn tail and flee, and be pursued by the second who in turn goes too far, etc.

If the intruder is a female, she bobs back a bit, is ignored or mated. She lays up to a dozen eggs on the ground, which hatch in due time when warmed by the sun.

Little brown lizards marked with five yellow stripes running the length of their bodies and with metallic blue tails attracted our attention when we first went to Sapelo. A little later we noticed large brown lizards with big orange-red heads peering from the great trunks and lower branches of the live oaks. Both lizards were examples of the five-lined skink. Skinks are actually divided into three distinct species whose habits and habitats illustrate a general ecological principle. Closely related species of animals that occur together in one area have differences of behavior, physiology, etc., that tend to keep them from competing with each other.

Individuals of all three species start life as the little blue-tailed, five striped fellows. As they grow, the tails turn gray, the stripes are less marked, and the males develop orange-red heads. The three species are difficult to distinguish, but they live in distinctly different places. The five-lined skinks proper prefer damp areas on the ground. We found them living around the artesian wells and the sawdust piles remaining from logging activities. The southeastern five-lined skink prefers dry areas and can be caught in hot sandy places. The third species, the broad-headed skinks, like trees in damp forests and is the species that live in the oaks. The broad-headed skinks are the largest of the three. The males grow to be heavy, foot-long creatures whose swollen, fiercely-colored heads have earned them the reputation of being poisonous beasts. It is startling to be confronted at eye level by one of these skinks as you round a tree; but actually they are harmless cowards, and they climb away as fast as possible.

In the fields and at the edges of the woods, one

TOP: *Five-lined Skink,* BOTTOM: *Six-lined Race-runner*

may encounter the six-lined race-runner, a lizard patterned like the young skinks but without the blue tail and with six rather than five stripes. But try to count the stripes! The race-runners, by comparison, make the skinks look stocky, stodgy, and very heavy-footed. Lizards' eyes are excellent. Any motion sends them scurrying. One race-runner was clocked at eighteen mph as he ran along in front of an automobile. It's true that this doesn't break any records. An athlete can do better in a hundred-yard dash. But the lizard's speed relative to its size is greater, and the most important difference between man and lizard is in the start. A man gets going slowly; his muscles are probably about as efficient as those of the race-runner at the same temperature, but they are much larger. According to an old rule, strength is about proportional to cross sectional area (length squared), while weight is proportional to volume (length cubed). So as an animal gets bigger it gets heavier faster than it gets stronger. The race-runner has relatively less trouble getting started; a situation we illustrated many times by trying to dive and catch one only to rise out of the dust empty-handed. A successful hunting technique rests on temperature. Since lizards are cold blooded, they are inefficient when cold. Occasionally one will venture out on a spring or fall morning before the air temperature is high enough for it to be fully active, with the intention of sitting in the sun and warming up. Then our

massive naturalist has a chance, as long as he leaps
before the sun's rays remove his advantage.

Spiders exist in almost all places in the world
except the permanently frozen or continually dry
areas. On Sapelo, the kinds and numbers are abundant. Their omnipresent fangs have a powerful effect
on insect populations.

The silk spider cannot go unnoticed by anyone
who walks in the woods in the autumn before the
first breath of cold weather. The female spider, a
handsome orange-brown, easily spans four inches,
including her long legs which have muffs of black
fur at intervals. The male is so much smaller than his
mate that he almost completely solves the problem
of how to avoid awakening the feeding instincts
rather than the passion of his wife by simply being
too insignificant for her to consider.

Striking in appearance as the silk spider is, it is
the web which strikes one in the woods. Or rather
one strikes the web and nearly bounces off its tough
silk. Several of the spiders, one female and her several husbands, may appear to be living in one big
happy home, but only the female contributes appreciably to the net of sticky silk that is a terrible hazard
to even the strongest flying insects. The female builds
a web of irregularly sized and shaped meshes, which
she may suspend between trees thirty feet apart and
twenty feet or so from the ground. She does not spin
a new web every day or so, as do many weavers; in-

stead, she uses her time and energy in strengthening and enlarging her original effort.

The beautiful and poisonous black widow occurs on Sapelo, but we rarely saw one, and then only when searching in suitable dark, damp places.

One of the orb weavers on Sapelo builds a web in which the main supporting lines that form the triangle around the actual orb are decorated with a thicker silk at even intervals, making an attractive pattern visible for quite a distance. The spider is one of outstanding appearance, with a broad, cream colored abdomen, a rather square shape, and body corners drawn out into little bright red points.

A spider found in moist areas is the primitive purse spider, a relative of the tarantula. At the base of a tree the female builds a silken tube that rises upward; this she decorates with bits of bark and lichen. Then she sits at the bottom of the tube waiting patiently for an insect to alight on or walk over the upper portion. When one does, she dashes quickly up the tube, stabs the victim from beneath with her great jaws, and cuts a slit in the tube through which she draws her prey. The unwanted parts are later

Purse Spider

thrown out through the slit and the tube repaired.
It seems unbelievable that the purse spider manages
to obtain enough food in this way, until one remem-
bers the hordes of insects that are forever crawling
about at ground level on Sapelo.

We were driving along a road one night in summer when we noticed many little spark-like reflections of the headlights at intervals on the road. At
first we thought they were reflections from particles
in the paving, but we discovered they were reflections
from the eyes of hundreds of wolf spiders roaming
in the night. One large female wolf spider carrying
her brood, enough tiny spiders that they resembled
fur on her back, reared up to do battle, undaunted
by the giants she faced. She gave up her threatening
posture when we weren't intimidated and sent the
baby spiders spilling off her back in all directions,
a mass exodus from danger.

Velvet ants with their bright orange-and-red furry
bodies stalk over the ground like ludicrous blood-
hounds tracking a scent; for their posture is unusual,
rear ends up in the air on tall back legs, heads close
to the ground over short front legs.

Velvet ants are a kind of wingless wasp; but everywhere there are real ants. Red pepper ants, carpenter
ants, small black ants, big brown ants. Ants are
surely one of nature's most successful creations. We
can make a sweeping statement like that and hardly
feel the need to justify it, since ants are so abundant

that even apartment dwellers in large cities meet them in their homes. Certainly no one could sit down on the grass on Sapelo for a minute, much less live on the island all the time, without being aware of ants.

Ants are particularly abundant for two main reasons: their social structure and their intelligence. The social insects, bees, wasps, ants and termites, are able to cooperate in performing tasks which a single insect couldn't accomplish. Although moving large pieces of food or killing large insects are impressive feats, the most important social function may be brood care, which reduces to a minimum the very great hazards of trying to grow up as an insect. An important aspect of this care is the control of the environment of the larvae. Here ants excel even among social insects, since they can move their young to various nurseries as their needs change in regard to temperature, humidity, and the softness of the ground on which they lie. The last is important to larvae about to spin the cocoons in which they will change to adults.

The intelligence of ants is minimal compared to mammals, but for insects it is astonishing. Ants can vary their behavior to suit conditions. Not only do they possess instincts to suit most needs, but they can also learn and modify their actions. And in spite of long working hours, individual ants may live for several years, long enough to put their learning to use, an advantage certainly, especially when this is compared with the few weeks of a worker bee's life. Cicadas live 17 years, queen termites may live to be over 25, but these are both such sedentary types that they have almost no experiences from which they could learn even if they were able. The ants really have no peers among their insect relatives.

Some kind of Sapelo ant is found in every part of the island: carpenter ants, which don't eat wood like termites but excavate it to make their galleries; pharaoh's ant; and the little reddish house ant that is indiscriminate in invading kitchens across the land. Another species that gets into houses but that northerners are well free of is the fire ant, named for the quality of its sting. Pick up a scrap of food from the floor that this small ant has discovered before you, and your hand is fiercely bitten and stung with a pain all out of proportion to the size of the creature. Fire ants sometimes destroy entire clutches of ground-nesting birds by entering the eggs as they are pipped by the hatching chicks and stinging the chicks to death.

Many animals seem to avoid ants although they eat other insects. Still, with so many ants available, it is inevitable that they will be eaten, and indeed they are. Flickers make a specialty of them. Lizards enjoy them. An enemy which takes a great toll is the ant lion. This insect is highly specialized to make the most of the abundance of ants. For the ant lion, the sandy, often dry soil of Sapelo is ideal. The larval ant lion lies in wait, buried in the soil at the bottom of a conical pit it has dug. A scurrying ant walks along the crest of the depression and is carried down into the bottom by loose rolling sand. There is a brief struggle; and the ant disappears under the sand, secured on the fearsome jaws of the waiting lion.

Hornets, wasps, and bees fly their ways, doing their jobs almost unnoticed in the midst of the bigger, easier to watch animals. Praying mantises stalk the woods to satisfy their ravenous appetites. Bright green and blue beetles line plant stems. Huge unidentified (by us) beetles occasionally stumble across the paths. Always in the evening there is the din of the cicada and the blundering of June bugs against the screens.

The Sapelo year might be divided not only into summer, winter, spring, fall, but also into sand fly season, mosquito season and green-head season; it was this division of seasons that controlled life on the sea islands during Plantation days. Now we cover ourselves with insect repellents, drain swamps, even send around fogging machines that envelop trees and marshes with clouds of D.D.T. mixed with oil; but in the 1800's there was little protection against noxious insects except skill in swatting. The plantation owners had an island house and a mainland house. It is often thought that the women and children were sent to the mainland to escape the sweltering heat of the humid islands, but the primary consideration was to get away from mosquitoes.

The mosquitoes breed in the marshes by the millions, and since the breeding ground cannot be eradicated, the adult insects must be conquered. Many types of mosquitoes are common, and many more are present but fairly uncommon. Without knowing too much about mosquitoes, anyone could pick out four types. First, there are the common house mosquitoes, and in warm wet periods these are more than common. The white stucco outside walls of our house were grayed in places with a thick covering of ravenous female mosquitoes. It is easy to distinguish the males, aside from the fact that they don't bite, by their large, rather attractive, bottle-brush antennae.

Another abundant mosquito is one that breeds in the brackish pools that form in the salt marsh next to the land after heavy rains. These are especially abundant just after the rain squall line moves over Sapelo in early autumn. Driving and walking through the landward parts of the marsh is impossible then. Once our vision through the windshield was noticeably reduced by the clouds of mosquitoes stirred into action by the jeep rudely running through their homeland.

A third type is noticeable because it is about twice the size of the common house mosquito. It is rather rare, fortunately, for its bite is about twice as painful as that of the common mosquito.

It was something of a shock to see the fourth type of mosquito, for instead of being delicately poised on four legs, it stood on its head when it bit, proclaiming itself to be an *Anopheles,* the malaria carrier.

When the plantation owners went to their summer homes to escape the swamp fever carried by the *Anopheles,* they left their slaves behind, who fortunately had achieved a relative immunity after centuries of adaptation in malarial regions. The disease known as sickle-cell anemia is related to this immunity. Individuals who inherit a gene for this from only one parent have a marked resistance to malaria, but individuals whose parents both contributed a gene for sickle-cell anemia develop the disease. The abundance of the gene is thus controlled by the bal-

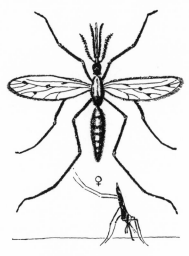

Anopheles Mosquito

ance between the benefits of resistance to malaria in some and the harm, usually death in childhood, caused by the anemia in others. About the same proportion of Negroes brought to Georgia probably had the sickle gene as their descendants now have in areas where malaria is severe, about twenty percent. Only about nine percent of the American Negroes now have the sickle gene since the absence of malaria removes the benefit formerly conferred by the gene and leaves only the harm.

Two more noxious pests are ticks and roaches. Suffice it to say that these abound on Sapelo.

Amphibians

FROGS AND TOADS are found all over Sapelo, from the fresh-water areas on the back beaches to the high oak forests. One afternoon, in what must have been a particularly suitable year, there were literally thousands of tiny, dark brown oak toads visible, hopping about on the floor of the mixed-pine forests. We wandered through them for some several miles, at which point we gave up trying to find an end to them. Within a few days they had apparently disappeared. Most had, no doubt, ended as dinner for a host of birds, lizards, spiders, and snakes.

One beautiful, balmy spring night we went out amphibian watching; in woodland and pasture ponds we tried to find as many different kinds of frogs and toads as possible. With the aid of a flashlight, it is possible to locate the frog and toad singers and watch them fill their throat sacs with air and pour their love songs into the night. During the hunt we found eleven species, from the southern bull or pig frog, six inches long in the body, to the half-inch long grass frog. The little grass frog, so delicate one can nearly see through his body, is a member of the tree frog or tree toad family, creatures which have adhesive disks on their toes to help them climb. We first heard their tiny, high-pitched calls coming from the grass in a roadside drainage ditch. Name any frustrating experience, and the finding of this frog equals it. We crouched as close as we dared to the ditch from which the noise was coming, pointed the flashlight and turned it on. The chirping continued for a moment and then stopped, and we saw nothing but grass. We turned off the light, moved to another spot, wildly brushing off mosquitoes, and waited for the sound to begin again. This was repeated again and again; and it was only later, after we had given up and then returned, that we saw, not at water level where one naturally looks for frogs, but up on the leaves of grass, frogs so tiny that their weight didn't bend the grass. This smallest of North American frogs uses grass as a substitute for trees, and climbs up perhaps fifty body lengths, that is two feet, from the ground.

On several occasions we came upon small forest

ponds that we could not find again in the daylight because the frog chorus that we followed to these out-of-the-way places was silent when the sun was up.

The tree frogs, taken together, are much more numerous than the typical pond frogs; and it was the former that made most of the noise. After a rain on a warm night we were awed at the thought that so much noise could be produced by such tiny creatures.

A large pond or marsh well populated with love-sick frogs is filled with a deafening cacophany. When near, we could actually feel the sound all over our bodies. The intensity can only be felt at close range; it drops rather quickly with distance because of the fast absorption of high-pitched sounds.

The call of the frogs needs to be loud enough to be heard at distances relative to the size and distribution of the species; for the call serves to bring together the members of a species at a spot suitable for mating and egg-laying. Small and relatively abundant frogs need calls with less carrying power than the large and rare or highly dispersed species. It is well that things work out this way, for a small species, other things being equal, will have smaller sound producing apparatus and so produce higher pitched, more rapidly absorbed calls.

The green tree frog is the most abundant of the abundant tree frogs and could be found throughout the woods and even in the houses on wet evenings. A few could always be heard calling from the live oaks at any time of year when the weather was damp and warm. From this habit of calling in moist periods, they have also come to be called rain frogs. These, with the pine woods tree frog and the squirrel tree frog, accounted for most of the chorus members.

To the somewhat harsh and unmusical racket of the tree frogs, the southern toad adds a musical, high-pitched trill. In some areas there are enough narrow-mouthed toads that their little buzzer-like sound is noticeable as well. These plump amphibians, with pointed heads, are extremely secretive creatures and hide themselves in the grass.

Most of the true frogs live in ponds other than those with large populations of tree frogs. They seemed to prefer the deeper ponds and drainage ditches, but we also found leopard frogs in a grassy pond in the cypress swamp. We traveled by jeep to this lonely spot, far from habitation, visited only by lovers of frog songs and frogs. The jeep bumped over two wheel ruts of logging trails through palmetto thickets and surged through grassy woods un-

Leopard Frog

der a foot of water. Finally, we had to get out and walk in hip boots into a murky and rancid smelling swamp—physical properties not usually associated with romance, but the leopard frog seemed not to know this. There was no other sound to be heard in the place but his love song.

Frogs and toads breed in a great many kinds of ponds, from rain pools in the pastures to permanent fresh-water ponds and swamps. There is something of a problem for those frogs growing up in the temporary pools. They must develop from egg through tadpole to little frog before the water dries up. It isn't surprising to find that the true toads, which live farthest from water and often use temporary ponds, can make it from egg to frog in a little under three weeks in warm weather. The narrow-mouthed toads take just over three weeks to develop, and the tree frog requires about five weeks to complete the cycle. The latter two species are apt to choose breeding places in the woods rather than in open fields, and so are better sheltered from the drying effects of wind and sun. The true frogs stay close to the water. They may wander short distances in moist areas (e.g., the leopard frog) or almost never leave the water (e.g., the pig bullfrog). These frogs require several months to pass through the tadpole stage alone.

We never came cross a terrestrial salamander, although we searched. We concluded that it was another animal that should logically have been on Sapelo but that in some way did not make the crossing from the mainland.

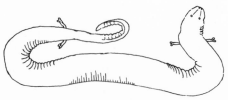

Aquatic Salamander

There is at least one of the aquatic salamanders present in the permanently-water-filled ditches on the island, though. We saw them frequently but never had one in hand. As we stood hipboot deep in the water of a drainage ditch choked with cattail and grasses, we could easily spot the dark, mud-colored salamanders, which ranged from eighteen to twenty-four inches long. When a salamander poked his head above water, we tried a fast dip with a net or a respectful grab by hand. But always we were foiled by the depth of the water, the difficulty of moving either net or body through the grasses, the easily stirred up mud, and the knowledge that this big salamander bites. No salamander is poisonous, although many are thought to be; but a long, slippery, slimy, squirming animal with a savage bite doesn't have to be poisonous to be respected.

These salamanders provide one half of a nice example of one of the simplest interactions between species—eating each other. Where the aquatic salamanders were abundant, the aquatic frogs were rare. We can't prove the connection, but it seems likely that a hungry salamander slithering quietly along in the mud at night would be very hard on big, plump pig frog tadpoles resting there.

We have named only a few of the thousands of kinds of land animals making use of every available kind of food, occupying every kind of dwelling. The sea islands have a very complex nature and exhibit interactions between species that would take years to study: how many of one animal are actually killed by another; what sorts of plants can serve as food or homes for a particular animal; how does the competition between two species for food affect their distribution, etc.

In the marshes and on the beaches, it is easier to see the relationships between animals because these habitats are more uniform and restrictive. But even there such studies take time. Ever since the island was raised from the sea, animals have filled every space in the functional network of who eats whom, when and where. A new invader would have a hard time finding an empty niche. He might make room for himself by forcing an already present species to move over a bit. Some of the mainland ground squirrels or mice could push the moles and rice rats out of spaces that the latter don't really find ideal anyway, the dunes for instance, and thereby make a niche for themselves. But as each animal and plant is restricted more and more to its most suitable environment by continued invasion of new species, this pushing aside of established species becomes more and more difficult. Finally, unless the climate changes or man makes major alterations in the habitat, the invasion will be said to be essentially complete and only species to which evolution gives new inventions that give them a decided advantage over the old guard will successfully establish themselves. Sapelo, as we have indicated, is approaching, but hasn't reached, this final condition.

Fresh-Water Ponds—
The Bird Colonies

TOWARD THE NORTH END of the island lie two
artificial ponds, made by bulldozing out and diking
up a low, marshy area. The ponds resemble a sea
of cattail plants dotted through with bits of open
water. Vegetation around these ponds is especially
lush. Great vines hanging from every available tree
or shrub make the area almost impenetrable.
Guarded by forests of oak and tall pines and by large
tracts of palmetto, and having the advantage of being at the most distant point on the island from habitation, the ponds have attracted vast numbers of
birds that prefer to live or nest in private—not in
individual privacy, however, as a nesting group may
number in the hundreds or thousands.

75

Even in winter, birds live there in great numbers. Flocks of ducks pause for several months during migration or stay throughout the winter—mallards, gadwalls, pintails, and teal; canvas-backs, ring-necks, scaup, and ruddys—the former group, puddle ducks, up-ending in the cattail shallows; the latter, divers, congregating in the deeper open-water areas. Floating with the ducks are common gallinules, grebes, and, in greater numbers than anything else, coots.

In the fall, before it gets too cold, the alligators are active. During the period of still warm days but cool nights, we saw our only example of an alligator catching food. We were watching a flock of canvas-backs floating in a small open-water area. Suddenly one bird began to squawk and thrash madly about as it was pulled under water with considerable effort, apparently by a small alligator. The other ducks seemed to take no notice of the disturbance, but several other alligators made an appearance and headed for the spot where the bird had disappeared. If they fought over the unlucky victim, it was entirely under water. No other agitation was visible.

Alligators are quite common on the ponds. We would see several sunning themselves on warm spring or fall days. They are remarkably fast on land, and when we surprised them, they would run for a water hiding place at such high speed that we were always glad not to be in the way.

Alligators live in fresh water but must have crossed salt water to reach Sapelo. We did find a small one living for the summer months in a tidal creek. Most individuals stay away from the sea and confine their wanderings from fresh-water to fresh-water areas. Two aspects of alligator family life came to our very close attention. One spring a male lived so near our apartment that we could hear his powerful roaring in the late evening, as he advertised for a bride. Another time we came across a number of young that had only recently hatched. Dark markings showed up starkly against a rather fresh yellow on the rest of the body, quite unlike the rather uniform mud color of the adult. Since the mother alligator was probably close by and since the young made considerable noise with their ferocious hissing, we didn't stay to observe them but took one home with us as a house guest. His bravery in no way matched his size. For the three days he was with us, he was willing to take on everyone and anything of any size. We jeeped him back to his family. It is said that alligators sometimes remain with their mothers into their second year, a very unusual situation in the reptile world, where the mothers usually abandon the future generation as soon as the eggs are laid.

The alligators are very much a part of the pond life, but their concentration there would not be so great if it were not for the birds. These supply considerable food; young that lose their foothold while climbing in the trees over the water and drown, and unlucky or ailing adults that are caught or die and fall into the water.

Sometimes in April, the ponds begin to take on a heavily populated look as the heron nesting season begins. Great, white, common egrets (egrets are herons, too) begin construction of their sloppy twig nests in low trees and shrubs on the islets in the ponds. They then begin their courting, in which they display their fancy plumes, which once almost led to their extinction. The other major plume-hunter's victim, the snowy egret, along with green herons, Louisiana herons, little blue herons, black-crowned night herons, and occasionally a cattle egret, begin nesting soon after the common egrets.

Redwings, boat-tailed grackles, anhingas, common and purple gallinules, and white ibis also nest on and around the ponds. Of all the swimming and wading birds breeding on Sapelo, only the great blue herons are standoffish and prefer to nest in a colony by themselves. Their nests are clustered at the tops of the tallest pines in an isolated part of the island.

During the height of the heron nesting season, it

Common Egrets

seems that bird is piled on bird, and nest on nest, on the pond islets. The din is terrific, and the special smell not less so.

All heron nests are similar: loose masses of twigs poorly put together. Invariably some are so poorly made that they collapse in storms or from failure of the thin supporting branches to sustain the weight of adult birds and eggs or young. After a disaster, when the eggs or young have been unceremoniously dumped onto the ground or into the water, the parents usually begin the nest again without profiting by their errors and with no more caution than before.

We undertook a study of the heron nesting one summer and visited the ponds every Sunday early in the morning before the sun was high enough to cook the eggs or injure the young while their parents took flight. There is no need to worry about the eggs cooling too much! The parent bird sits on the eggs to keep them warm only at night; in the day it is to protect them from heat. We marked the nests, weighed and counted eggs and young, and tried to determine what was responsible for destroyed nests by examining the remains.

We began by marking nearly fifty nests of common egrets, and wanted to do the same for all species found on the islets. Our hands quickly got too full with the first group to manage as many of the others. We started rather placidly. We numbered the branch or trunk near a nest with red paint, then weighed and counted the eggs in a nest, usually two to four. The parent birds flapped away through the shrub tops; some flew around overhead, but most of them just flew from tree to tree as we moved along.

The chicks hatched in three to four weeks. Baby herons are somewhat intermediate between the completely helpless and naked song-bird young and the active, downy young of turkeys, gulls, and shorebirds. Young herons have a scraggly down covering, but can move very little after hatching. It would be dangerous for them to do so because they might fall or damage the nest. They sit still, trying to look as unobtrusive as possible and failing completely. The little balls of green skin covered by the moth-eaten-looking white fuzz stand out clearly against the leaves and branches, especially since the nests must be exposed to allow the large adults a clear landing.

Nearly half of the common-egret eggs were eaten by crows and raccoons before they hatched, usually soon after being laid. The nests are often left unattended for short periods, especially before the full clutch of three to four eggs is complete and immediately after. The female lays no more than one egg a day. There is a period of a week or so then when the large, pale blue eggs lying in the conspicuous nests are highly vulnerable. During later stages of incubation and after hatching, one parent is always at the nest, and so predation becomes slight. As the young grow, they are able to protect themselves and are again left alone when both parents are busy fishing. At this stage the nestlings begin to climb about on the branches and twigs near their nests, using wings, beak, and claws to help themselves. Since their feathers have only just begun to appear, they remind one of a puppet skeleton animated by a child. As they grew more expert and we could no longer catch them, we counted from a distance and gave up weighing. At this age, almost fully feathered but yet unable to fly, the birds began to wade into the cattails, perhaps to try fishing for themselves. In the nests we observed, only forty-five percent of the eggs survived to grow to adulthood.

Most of the nests studied were located on one islet heronry, which held over one thousand nests in an area about the size of an average suburban front yard. A stroll under the trees was accompanied by the pleasant-sounding spatter of falling rain, even on cloudless days. We wore wide-brimmed hats and clothes reserved for such trips and learned not to look up. The grayish-white guano covered everything on the islet. The topmost branches on the shrubs were the customary brown-green, but the lower branches wore a chalky, gray-white cover.

The young birds have very unpleasant personal habits, which are accentuated when the birds are disturbed. The most pugnacious creatures were the young black-crowned night herons. A weighing sequence usually went like this: As we approached the nest, the young vomited their last meal with considerable vigor and good aim. Since the meal might have consisted of a nine-inch mullet, the intruder was dealt quite a blow. Because the usual intruder is looking for a meal anyway, the young bird uses good strategy in offering a substitute, by hitting him with a chunk of food even if it is slightly used.

We weren't looking for food; and although wary of falling fish, we were not deterred by them. Seeing this, the young herons would abruptly turn around and empty the other end of their digestive tracts in our direction. Next, they would defend themselves by opening their beaks to an unbelievable extent, squawking loudly, and finally thrusting the beaks in the general direction of the nearest human head. The young of other species defend themselves in similar ways, but usually less effectively.

There are always a few white ibis nesting among the herons. During most of the years we were on Sapelo, only a handful were counted nesting. These were usually seen late in the season, which may indicate that they had tried somewhere else and had been unsuccessful. The ibis are not like the herons in their devotion to a nesting site. The colony often shifts about. In the summer of 1958, when we made the nesting study, a colony that usually nests in South Carolina was missing and over a thousand white ibis took to the trees on the islets. Presumably the Carolina group was trying Georgia.

The white ibis nest is somewhat better constructed than that of the herons, more cup-shaped and usually lined with fresh green leaves. Two or three dark cream-colored eggs mottled with brown, slightly larger than hens' eggs, are laid. The ibis just after hatching are no more attractive than most other young birds that remain in their nests and, in contrast to the herons, have their eyes shut. They grow quickly on a diet that consists principally of small shrimp, a species that does not grow large enough to be commercially important. In a week to ten days, they can leave their nests and climb about in the trees. By the end of three weeks, they spend most of their time in the topmost branches begging food from adults.

The arrival of the ibis in the heronry filled the trees with nests. Most of the leaves disappeared, pulled off for nest lining or accidentally knocked off in the general hubbub. A few weeks later the trees were loaded with young birds climbing by means of beaks and feet.

Of the nests we studied, the black-crowned night herons proved to be the most successful parents. In

White Ibis

all their nests, young were successfully raised, except in one where the parents disappeared and the young died. The other species raised at least one young in only thirty-five to fifty percent of their nests. On the basis of individual eggs rather than nests, we found that the larger common egrets and night herons successfully hatched and raised from one-half to three-fourths of their eggs, but the smaller species raised only one-forth to one-third of theirs. Most of the larger birds' losses were in the form of eggs; more young than eggs were lost by the smaller herons. The nests of the big birds are exposed and the eggs conspicuous, but the young are large and able to defend themselves. The nests of the smaller birds are hidden and don't attract the attention of predators until the relatively helpless young are moving about in the nests.

We developed a real interest in the young birds, despite their bad personal habits, and felt a loss when any disappeared. A nest full of healthy young one week might, on the next week's visit, be represented by only a few broken branches and scattered sticks, telling a story of heron tragedy and raccoon dinner.

Not all young heron deaths can be attributed to predation. In many nests, the last young to hatch would grow more and more slowly in comparison with his nestmates and would eventually starve, although the others were healthy and well fed. The older siblings are better able to attract their parents' attention and get food for themselves. The youngest, on the contrary, doesn't get enough to grow properly, and so it gets weaker and weaker until it is completely unable to beg for food and starves. This, as is the case with owls, is nature's way of enabling the birds to take advantage of an occasional extraordinarily good supply of food. If each heron pair lays two eggs, enough young will be raised and survive till the following nesting season to permit the population to maintain itself. But there will be no extra survivors, on the average, to increase the population. Should the supply of food increase, the heron population couldn't use the additional food, and another animal would be able to pick up the extra tidbits. However, if each heron pair routinely lays an extra egg and hatches a third chick, a chick destined to starve in ordinary years, there will always be an extra mouth ready to fill with any extra food the adults can find. Thus the heron population will grow to the largest size that the available food supply can support.

There is such a heavy concentration of birds in the heronry, that there is no possibility of the immediate area providing enough food for parents and babies; but herons are large birds and range widely. Colonial nesting must provide advantages, perhaps mutual stimulation to breeding, that offset the disadvantage of not being able to fish right under the nest.

The egrets fish in the estuaries and creeks in the

marshes. The ibis probe for shrimp in the marsh edge. Cattle egrets fly to the pastures, six to ten miles away from the nest sites, and return with their stomachs full of grasshoppers for the young.

Cattle egrets did not live on Sapelo when we arrived in 1955. They came on June 6, 1956. We know the exact date because we had been looking for them. This somewhat dumpy looking egret has been expanding its range northward from South America, to which it came from its native Africa. For some years it had resided in Florida and had been seen in South Carolina. This meant it had either been unobserved or had flown straight over Georgia, and Georgia is no less desirable than South Carolina to cattle egrets. The logical place for the birds to be is walking along beside the legs of grazing cattle, where they pick up the grasshoppers and insects that the cows stir up out of the grass. We concentrated on watching the large cattle pastures near the south end of the island, where several hundred head were fenced in. Each time we passed, which was every day for one reason or another, we gave the pasture a sweep with the field glasses. Our patience paid off; for on June 6, some weeks after the main migration of egrets, the cattle egrets came, displaying full nuptial plummage, the buff-reddish patch of feathers on breast, head, and back.

The other egrets and herons had already taken to the sequestered heronry and were well on the way to parenthood. That first summer the cattle egret didn't try to catch up. Three birds came, ate in the pasture, roosted nearby in a dead tree, and flew south again in early September. In 1957, three arrived in May, and two more in June. Then one individual deserted the roost in the tree near the pasture and set up a nest in the top of a shrub near the open water of the duck ponds. Three eggs were laid by July. On our next visit, there were three more. Three chicks were successfully hatched and grew to adult size. They may have been among the ten cattle egrets that returned to Sapelo the following year. Unfortunately, we found no more nests of this bird in successive years, although the number seen among the cattle increased to twenty individuals.

An interesting but unrelated bird, the anhinga or water-turkey or snake-bird also nests around the ponds. A few pair build their equally unsound nests among the herons. As one of the alternate names suggests, the anhinga has a long neck at the end of which is a head only slightly larger around than the neck. The neck has a kink built into it so that it may be carried half-folded. The bird sits on the pond, and by varying the amount of air in his body, can sink slowly under water or rise as he wishes. He fishes by sinking into the water and pursuing a fish, which he spears with his sharp beak by suddenly unkinking his long neck. He comes to the surface, shakes the impaled fish from his beak, and takes it to his woolly, down-covered, large-footed young. His wing feathers get wet, and instead of shedding water like most other birds, he must spread his wings to be dried by the sun. He shares this unfortunate condition with his relative the cormorant; but no one knows why either bird should be so afflicted.

The nesting is nearly finished by September. The ibis fly south. The herons disperse to their winter feeding grounds. The raccoons and opossums go into the woods when there are no more eggs or chicks available. The snakes and alligators disappear, presumably into hibernation. The islets return to their uninhabited state. The twigy nests are blown out of the trees. The rains wash the droppings from the branches and ground. A second set of leaves comes forth to repair the damage caused by the multitude of birds. The ponds are quiet and peaceful again, for a time, until the noisy ducks come squawking in for the winter.

Anhinga

PART III

The Marsh

The Form of the Marsh

SAPELO ISLAND grows and shrinks two times each day. Like an expanding and contracting giant, it changes shape, now bulging and rising, then shrinking as a surge of water pours through or out an inlet. This fluctuation is due to the tides, sweeping in onto the salt marshes and then ebbing back into the nearby ocean. The tides make Sapelo an island. First it is set off by water on all sides, open ocean to the east, and marsh with just the tips of grass showing to the west; then six hours later it seems a part of the mainland, a bit of forest separated from the coastal plain forests by a broad grassy savannah.

When we live on land, we are conscious of two natural cycles that control our activities: first, the

91

alternations of day and night; and second, in temperate regions particularly, the seasonal changes—cold in winter, replaced by the beginning of growth in summer, and maturing and dying in autumn.

The tidal cycle overlies these daily and seasonal changes and may greatly modify their effects on the inhabitants of the tidal region. A high tide may give protection from either the heat of a July sun or the freezing of an unusually cold February night. Or a low tide can expose a normally salty area to a thorough dousing by a summer thundershower. On a sunny March day an animal living in the unshaded mud of a tidal creek bank may be exposed to a change in temperature unlike any on land. For example, on one particular day, at high water a few hours after midnight, the mud was a comfortable (for a worm) 50° F; but as the tide ebbed and a slight predawn breeze sprang up, the cold air plus the evaporation of water lowered the temperature to 30° F. The sun rose and shone on the black mud, unprotected by plants, and heated it until by eleven o'clock the temperature was 80° F. The flooding water reduced it abruptly to 50° F an hour and a half later; and, since the sun that afternoon was too low to provide appreciable heat after the sea receded, after the tide went out, the temperature dropped again to near 32° F. A number of organisms manage to live under these rigorous changes, but certainly their powers of adaptation must be tried to the limit.

Marshes don't exist exposed to ocean surf, for the pounding of wave after wave would wash the soft mud and plants away; but marshes are always, and strongly, affected by the flushing tides. The tide flows into the marsh with a relatively gentle action. Then there is the almost magical moment when it turns to flow out again. The local people say that the wind dies just as the tide changes, and we've noticed this several times while working in the marsh. But we'd be unwilling to say there is any connection. This is the sort of phenomenon one notices when it happens, but a failure of the wind to change, to stop cooling and to stop blowing insects away, is not so clearly marked.

During the high spring tides, the whole marsh of Sapelo is covered. The marsh plants are completely submerged and then are thrust up six hours later into the air. The creeks, broad, deep waterways at high tide, lapse into thin trickles surrounded by extensive mud banks at low tide.

The winds further complicate the involved picture of the tides. A strong onshore wind may pile in extra water, which boils through the sounds between the sea islands, rushes up the tidal creeks, and flows over onto the marshes. This often catches unaware the non-aquatic animals that are living as close to the border line of water as they dare. Suddenly they are covered, and they either have to be able to withstand seawater or perish. An extra-high tide with onshore winds may mean disaster for the marsh wrens, which nest far out in the tall grass marsh along the creeks. Normally the nests are above water; and when the eggs and young become submerged, they are killed. Such a strong wind can also prevent normal low tide. This becomes a day, as the fishermen say, "when the tide didn't go out."

The extreme conditions that the interaction of the tidal cycle with the weather produce, the extra hot or cold day or hour, generally determine whether or not a species can exist in a region. For example, an especially low tide at midday under an intense July sun can have a disastrous effect on animals living low in the tidal zone that have no special resistance to heat or drying. As a result the marsh contains a rather unique set of creatures, living on and in its surface, all especially suitable to the specialized conditions or adapted to them. The raccoons and otters and other land animals that invade the marsh may seem important because they are of a size to be easily seen, but their numbers and effects are small compared to the millions and millions of crabs, worms and molluscs specialized for life in the marsh.

The tremendous amounts of water coming and going every twelve hours give the marsh its form. From the point of view of the water or of an animal floating in it, the water enters between the sea islands, goes into the sounds which may be a mile or more across, into large tidal rivers and then into progressively smaller rivers and creeks until it comes onto the marsh at the beginning of a tiny creek or spills over a larger creek bank. The flow is reversed with ebb tide; and except for rainfall on the marsh, the tidal waters are the only waters the rivers and creeks of the marsh carry. The *only* doesn't mean that the amount of water is small. Duplin River, running through the main Sapelo marsh, carries water that covers from five to six square miles of marsh to a depth of two to three feet on the highest tides. Most of this flows in or out during the middle three hours between high and low tide. Very strong currents are the result. The strongest part of the current races along the center in straight parts of the river, but touches and strongly erodes the outer banks where the river curves.

The marsh looks uniform viewed from the land, only broken up by creeks and rivers, but it has several parts that can be distinguished. If one started on the land and walked out onto the marsh, a distinct sequence of forms would be obvious. First, one would see a group of land plants, marsh aster and tough shrubby marsh elder, which grows at the edge of the marsh because it can withstand an occasional immersion in saltwater. Then, the marsh proper would begin, with the growth of *Salicornia*. This low succulent looks like a small brush that has lost all its bristles; its stems are swollen and green with water. In the fall this plant turns red, and the marsh where it grows takes on a characteristic reddish hue.

The marsh grass, *Spartina alterniflora,* also begins to grow here, with plants five to ten inches high. Further into the marsh, other plants disappear completely and *Spartina* is in sole possession. *Spartina* looks like and is a coarse, green grass. It varies tremendously in size and abundance, depending on the soil, and has adapted itself to a variety of conditions.

The vegetation here at the beginning of the marsh is very short. The plants are not usually close together. You can put your foot down between them. You may even walk some distance over patches of bare sand. But then the going gets rougher. You have to put on sneakers to protect your feet from the sharp leaves of the *Spartina,* which has become thicker and grows up to two feet tall. Slowly the *Salicornia* completely disappears. The ground becomes uniformly muddy. Ooze comes up around your shoes as you stand on the continuous mat of *Spartina* roots just under the surface of the marsh. The mat is only six inches to a foot thick, but it supports weight and acts underfoot like a very stiff

SALICORNIA

trampoline. A slight vibration can be felt when one walks. With a little spadework through the root mat, soupy mud is exposed underneath that supports weight only as quicksand or water does.

Around Sapelo there are large areas of this type of marsh, which may continue, acre on acre, without any obvious variation in the marsh grass. Then suddenly there will be a slight rise in the ground level as one approaches a creek or river. On the rise you sink knee-deep into the mud, especially if there have been spring tides or heavy rains because although the grass continues to cover the ground, the roots are no longer matted together. Along the banks of all the tidal channels this rise in ground level or natural levee is formed by the water when it overflows into the marsh. As long as the water is confined in the relatively narrow channels, the flow is swift and can carry with it the load of silt and clay brought into the estuaries by the rivers and picked up from the creek beds and eroding banks. But when this rush of confined water suddenly floods over into the much larger area of the marsh, its speed is sharply reduced. The dense growth of *Spartina* along the

SPARTINA ALTERNIFLORA

creek banks aids in slowing the water. The result is
that the silt and clay settle out, and a levee builds
up.

Spartina grows taller on the natural levee, to
about three feet, and since it is on higher ground as
well, it rises well above the *Spartina* on the lower
marsh at the backside of the levee. In places the
levee is high enough that during neap tides it is continuously above water. If it happens that a period of
hot dry weather coincides with a neap tide series,
the highest such places may dry out to some depth.
Then the oxygen in the air gets to the hydrogen
sulfide under the surface and changes it chemically
to sulphuric acid, which kills the grass and makes it
possible for erosion to reduce the levee again.

As one goes over the top of the levee and walks
down the bank to the creek, the grass gets progressively taller so that the tips of the leaves grow to
about the same height as those on the levee top. The
streamside grass commonly reaches nine feet in
height, and the plants grow very close together. This
lush growth of *Spartina* completely hides the mud.
The area looks bright green in contrast to the backside of the levees, which are the brown-green of mud
showing through between the more widely spaced
plants.

This growth of grass to extreme heights on the
creek banks gives the illusion that the marsh is uncommonly flat, without relief; but underneath the

Spartina, the ground climbs and dips and then levels off, as one goes from creek to marsh upland.

The Sapelo salt marshes have a half-marine, half-soil smell; part decaying vegetation, and part the smell associated with salt, fish, and iodine. It is not an unpleasant odor, even when sniffed from a raft of decaying *Spartina* stalks under a hot summer sun.

The salt marsh seems permanent, and *Spartina* is the backbone and strength of the marsh. But even *Spartina,* in spite of being hardy and adaptable, can stand only so much time under water. The lower parts of the creek banks, where the water flows for long periods during each tide, are bare of all plants except algae, which at times are abundant enough to give the mud the rich golden color of diatoms or the greens of other single-celled plants. The creek banks, since there are no roots or plant parts there at all, and since they are always waterlogged, are extremely soft. It wouldn't matter if you changed sneakers for hip boots—even these would be inadequate. On the creek banks you sink waist-deep in the brown mud—truly immersed in Sapelo Island.

Out With the Tide

WHAT IS IT LIKE, a world half-land and half-water? We often tried to get a better understanding of the marsh by spending time walking or floating about and looking at things, trying to be unnoticed by a muddy raccoon or using binoculars to get a close look at the normal lives of fiddler crabs.

The actual observations here cover four years, but we are going to describe them like a time-lapse film, compressing time so that the events of many days can be seen in one day.

We bipeds, used to solid ground under our feet, air all about us, and only the changes of seasons and day and night to concern us, have gone out to spend a day in the marsh.

100

In early morning, with the smell of the new day in the air but with no morning light yet showing, we have set out with a canoe. We, although having invaded the marsh only very lately, have already adjusted ourselves to its physiognomy. With a canoe we can go almost everywhere, even through the thin tricklets of creeks deep in the marsh interior.

The morning has been calm. As we've pushed off from the dock and quietly paddled along the shore of the Duplin River, we have heard the herons beginning their muted morning squawks in the restless period between sleep and wakefulness. Here and there we have heard the rush of flapping wings warming up for the morning flight, but then the stretching birds have settled down again for a quick nap before leaving the shelter of the night roost.

It has been cool during the mid-spring night, and most of the important marsh animals, the fiddler crabs and snails, have not been active. The tide is half-low and ebbing. Bits of marsh grass, pulled away from the banks of the creeks, float by us on the way out to sea.

Light begins to increase until even we, with man's poor dim-light vision, begin to see. A few night herons are fishing along the river bank, and a few mammals are moving about, concerned with the business of their predawn feeding. The marsh is vast; the mammals visiting it few and wary. We slip along looking for signs of larger mammals, and in a mile of paddling we see two raccoons and one mink. Farther down the river, off in a creek, two otters slip down the bank to quietly swim away with hardly a ripple. Ahead of us, at the juncture where a creek joins the river, we catch glimpses of a group of porpoises rolling about and blowing in the high marsh grass, which grows almost to the edge of the water. They have been rooting in the soft mud, probably for crabs or mussels. They seem not only to be eating but sporting in the shallow water. Even though we have made very little noise, the porpoises have heard us and seen us; but they let us come quite close before they move off with characteristic ease and speed.

The dawn expands in a rosy hue over the marsh as we turn into a large creek leading into the marsh interior. A few dark clouds have moved off out of sight. The night herons fly away to roost. The raccoons, which have been working the stream sides close to land for crabs, run dog-fashion back to their trees to sleep. The ones that are caught far out in the marsh by the advancing light fashion matted nests of the living marsh grass and bed down for the day.

Then the day herons begin to appear. Files of white egrets make their way through the air, reflecting the rose of the sky. Solitary great blue herons wing their lazy way into the marsh and settle to fish in the creeks. Little blue and Louisiana herons fly over the marsh, looking for likely fishing spots. Long-billed clapper rails, or marsh hens as they are widely called, walk along the creek banks probing the mud with their strong bills, stalking crabs for breakfast.

It is still early enough in the year for the long-billed marsh wrens to be singing in the tall grass of the stream sides. We see the males flitting from grass clump to grass clump, each building a series of clumsy nests in his area of the bank, the territory he has established and will defend noisily from other male wrens. Each flies up and down his domain, pausing to sing at various spots to let other males know that this is off limits. With the same song, he lets the females know that they are welcome—a combination of incompatible messages, the secret being not what the song says, but how the listener interprets it. A female flies to the male we are watching and selects one of the better nests he has made. Then with considerable effort she fixes it up as a better home for her brood, or broods, since she may have more than one in the same nest.

The male has built his nests by bending and inter-weaving *Spartina* leaves until he has achieved a roughly spherical mass, hollow inside, with an entrance hole in one side. But this male, like all male marsh wrens, has been more interested in quantity than quality. His nests are so loosely constructed and so full of holes that no one of them would ever keep rain off baby birds, or for that matter hold eggs safely.

It was to one of these nests that the female flew. Her repair work consisted of adding more grass blades until she had a tight structure without openings, except for the small entrance hidden low on one side. She was thorough and even built a little shelf projecting into the nest from the lower side of the entrance to keep the young from falling out.

The clapper rails we see probing the mud banks do not have to fear for their eggs or young when spring tides flood the marsh; for their eggs are adapted to withstand immersion in salt water, and the young rails are downy and able to swim from the time of hatching. On the crests of extra high tides with an onshore wind, we have often seen groups of these secretive birds huddled in the few clumps of marsh grass not completely covered.

The rail nest is almost impossible to find in the tall marsh grass, even though you can hear the bird's cackle carrying across the stillness of the marsh. The brooding bird does not fly from the nest and reveal its location, but gets off quietly and slips unobtrusively through the thick grass without disturbing a blade. Then, if necessary, she takes wing some distance away. We once watched this procedure while looking at a nest that had been built so close to a small area of open water that it could be seen with binoculars from the opposite side. As we watched, a cruising crow settled near the nest for a breakfast of eggs.

A seaside sparrow flies past; then another comes and disappears in the short grass of the flat marsh. An occasional redwing sits on the strong *Spartina* plants near its nest. Now more and more we hear clapper rails, but always in the watery aspect of the marsh, swimming and walking about on the mud while they eat crabs and snails. The wrens, as abundant, emphasize the land aspect of the marsh, acting like land birds, flitting from one *Spartina* stalk to another and feeding on insects.

Cattle move along slowly in a herd, feeding on the marsh grass. The area of their choice has been almost denuded of grass, changing its normal character. It should be an area of knee-high, relatively close-packed plants, but the grazing and trampling have reduced it to a sparsely covered, muddy morass. The cows take to the marsh both for food and comfort. The unhampered breezes blow away the flies that they collect in the deep woods where they spend the night.

Land and shore birds are all around, feeding on marsh grasshoppers, other insects, and seeds. Shore birds can be seen congregating here and there in areas of short grass, resting, choosing the marsh, perhaps, to avoid the wind-blown sand of the beach.

The day begins to grow warm now. The sun is some forty-five degrees above the horizon. The tide has fallen almost to dead low. Literally millions of fiddler crabs, the marsh's most obviously abundant animal, are moving about.

We have passed large herds of red-backed or sand fiddlers, *Uca pugilator,* living in the sandy areas of the marsh close to the land and on the river banks where strong currents erode particles of mud. An adult male, with the one large claw from which the species gets its name, can sit comfortably on a silver dollar with only the large claw hanging over. The youngest are only twice as big as the head of a pin. The blue or mud fiddler, *Uca pugnax,* lives in the areas of the marsh where the ground is muddy. For some reason, neither of the fiddler crabs chooses the tall streamside marsh grass area, but two relatives do —the brown, square-backed crabs, *Sesarma cinereum* and the purple square-back, *Sesarma reticulatum.*

Fiddler crabs will eat almost anything that comes their way in the form of fish and animal material; but large bits of food are rare, and they are forced to rely on tiny particles of digestible food found in the mud, bits of partly decomposed *Spartina,* bacteria, and algae. With a regular pattern of motion, a fiddler scoops up a claw-full of mud and stuffs it into

Uca PUGILATOR, *or Sand Fiddler*

its mouth. The males rest their fiddle to one side and eat one-handedly; but the females eat busily with both hands. This two-fisted eating doesn't seem to give the females an advantage. They grow to be only about three-fourths the size of the male and don't sport the big fiddle. The mouthfuls of mud are not swallowed, but are sorted by the six pairs of mouth-parts. The outermost serve mainly to hold the mud in place while the next few sets sift out the large lumps and pass the fine material along for further processing by the digestive tract. After five to ten handfuls of mud have been sorted, the rejected material is spit neatly into one claw and placed to one side.

The herd of thousands of fiddler crabs that we watch at dead low tide, has eaten along, denuding the mud of the detritus (diatoms and algae, bits of marsh grass) on the surface. The entire surface is marked with tiny scrape marks, showing where the mud has been picked up, and little pellets of rejected material have been strewn behind as the crabs have advanced. When we make slapping noises on the mud with our canoe paddles, the fiddler crabs move off a short distance in a miniature stampede.

During the time of mating, which may last all summer, the males set up territories around their burrows, entry into which is denied all other males. They wave their fiddles to warn males off and attract females. Each species of fiddler crab has a set pattern

of display, of claw-waving and deep and shallow
knee bends. The red-backed fiddler makes a noise
by rapidly vibrating the large claw so that two of the
joints rub together. If a female is sufficiently taken
with a male's waving and bending, they retire together into the male's burrow. Sometime later the
female produces a mass of orange eggs, which she
carries about under her abdomen, as do other crabs,
until they hatch. At the time of hatching, the female
goes down and stands in the water. The baby crabs
emerge to spend the first part of their lives as larvae
in the hazy cloud of plankton floating in the sea.
When the crab larvae are ready to molt and enter the
stage where they settle down as tiny but recognizable
crabs, they wait until the tide carries them to a suitable spot before they take up their life as mud or
sand dwellers.

With such a chanceful life cycle, enormous numbers of crab larvae must be released. The larvae
cannot make any appreciable headway under their
own power and must depend on the currents to carry
them to a suitable marsh. Great numbers perish
when this fails to happen. Others fall prey to the
many animals that feed on plankton, from other
plankton to fish such as herring.

As we paddle on in the canoe, the noise of the
marsh comes to us from all sides. Even though we
are often unable to see the source of the noise, or
even accurately tell the direction it comes from, we

realize that there are large numbers of animals carrying out their activities. Now that the tide is dead
low, the water is not absorbing all of the sounds
from the mud. At intervals, the loud rattling call of
the marsh hen comes to us, and the singing of the
marsh wrens continues. But there are smaller noises,
too: the minute clicking of oysters and mussels clos-
ing their shells and the scurrying rustlings of feeding
fiddler crabs brushing through the marsh grass.

Punctuating the bird calls and rustlings of the
feeding fiddlers are loud pops made by a crustacean,
the snapping shrimp. This animal, which lives in the
mud banks near low-tide level, has a claw with a
catch that holds it open until the animal exerts
considerable force. Then it suddenly closes with a
loud pop. There is power enough in this snap to
break a glass dish. Perhaps it serves to protect the
shrimp in his muddy hole.

The popping continues as we go on, but the world
of the marsh subtly begins to change. The exposed
mud banks along the creeks, which have lain bare
in the hot summer sun, have begun to cool again.
We have been paddling along through the bigger
creeks, sometimes barely sliding over the bottom
mud, but now we notice that the canoe has become
more buoyant. It doesn't wallow but floats. The
feedings of the crabs and birds continue, and pops
are still heard. But the tide has turned.

The Tide Comes In

WE TURN THE CANOE into the smaller creeks with the advancing tide, beach it in a relatively firm spot, tie it to a mass of *Spartina,* and with the help of a plank, climb onto the marsh where we will be able to get a closer look at the creeping and crawling things that inhabit the grass and mud. Many of these marsh dwellers have built-in clocks that warn them of the approach of the tide, and they begin to take protective action before the water reaches them.

The salt-marsh grasshoppers, which live on the *Spartina* plants near the creek banks, have begun a slow climb up the stalks to get away from the rising water. Already our feet are wet too, but the water is warm and not uncomfortable.

110

In the sides of the main, central stems of the *Spartina* that grows on the levee, we have found small holes made by ants. Ants live in nests in these stems; and if the nests filled with water each tide, the colonies could not survive. The nests must have openings, so a watertight door of some sort is needed. And one is provided. Before the tide rises, all the outside, foraging ants return to the nest, and then a big-headed individual simply plugs up the hole with her head. She is confined to the nest for life since her head is too big to pass through the door.

We have seen a variety of spiders stringing their webs between the blades of marsh grass or climbing about catching their prey by stealth or by sudden jumps. Their activities are interrupted by the tide which destroys the webs. The spiders can withstand immersion in salt water, however, because of their tough skins and need not run from the incoming tide.

Part of the prey of the spiders is now flying about our heads trying to prey on us. Sand flies, tiny midges with big bites and green heads, big green-eyed flies with even bigger bites, lay their eggs in the marsh and live their larval life within the mud. As adults they feed on warm blooded animals, which they will leave the marsh to find if necessary.

Land—wet, muddy land, that has been created by low tide—is slowly being submerged as the water rises. Consider what this means to an animal adapted

for living on land and in air but that cannot escape
the water. The air which he breathes disappears entirely, and he must be able to extract some oxygen
from the water or be able to wait without air until
the tide ebbs. The ground on which he walks remains much the same, but it is more difficult to walk
because the water that surrounds him is more viscous
than the air and, of course, it is completely impossible to hop or jump. Anyone who has tried to run
in knee-deep water can imagine the animal's re-
strictions. Many animals avoid these problems and
move out of the water by climbing the *Spartina*
stems; and so we also avoid the water by returning
to our canoe.

The sun arrives directly overhead. It is hot, and
thunderheads can be seen building up over the mainland. Any trace of the larger mammals has vanished.
Raccoons remain asleep in their grass nests. The tide
has covered their footprints. The cattle have re-
treated before the rising water and sought shelter
from the heat in the woods, the biting of the flies
being the lesser of two evils.

The smaller animals, the insects, some kinds of
crabs, snails, and spiders have begun migrating up
onto the grass or downward into the mud. A great
section of level marsh looks like an orchard with
thousands of white snails swaying on the grass, like
so many blossoms. Grasshoppers and spiders cluster
on the grass tips, seeking shelter on a spot that will

remain terrestrial. The congregation on some leaf tips has become so dense that they are bent with the weight of the animals. Soon the birds, gulls, terns, and even land birds like the painted bunting and barn swallow, come out over the marsh to harvest insects. The grasshoppers remain quietly on the tips of leaves until they see something move near them, then they crawl down the stems to take their chances under water. Sometimes they make the mistake of jumping, which usually means landing in the water and being carried away by the current until they are eaten by fish or birds. The grasshoppers that hide out under water have a fair chance of surviving if they are not forced to remain under too long.

Many of the marsh animals can either breath normally under water like the crabs or, as with the spiders and smaller insects, enough air clings to their body surfaces so that they can survive under water for several hours. But even for those that can live under water, vision and movement is greatly restricted and their enemies, who were easy to spot in the clear air of the terrestrial marsh, in the shadowy depths of the aquatic marsh are much harder

Snails on tips of marsh grass

to see. And not only is it much harder for them to see and avoid the low-tide predators, but with the advancing tide, a whole new set of predators moves in looking for food.

Fingerling fish come right up onto the surface of the mud looking for insects and then wriggle back to the water. Larger fish, looking for fiddler crabs, come up to the edges of the grass along all the creeks. Mud crabs come out of their holes where they have been waiting out the low tide. Dark forms of blue crabs swim in to prey on other crabs, small fish, and worms. Even small squid invade the marsh, looking for a meal.

In the face of all these difficulties, the low-tide animals must be very resourceful and clever to protect themselves. Different species of crab have different means of protecting themselves at high tide, which account for their success in living in different parts of the marsh. The brown square-backed crabs climb the *Spartina* like the grasshoppers, but being heavier and not so agile as the insects, they are more apt to be knocked off. This disadvantage is offset by the fact that they live only in the streamside marsh, where the grass is thick and tall and also so close to the land that they can climb out of the marsh entirely. The purple square-backed crabs are considerably more pugnacious and risk meeting small fish. They are most active just as the tide is flooding the marsh. Before the water is deep enough to permit larger predators to swim in, the crabs retire to their burrows, which are more complex than those of other crabs, often having several entrances. This behavior provides a good example of the fact that closely related animals living in the same area, the fiddlers and square-backs in the marsh, tend to behave in ways that prevent them from competing. Whether the purple square-backs found it necessary to become pugnacious because they chose to be active as the tide rose or were able to be active then because they were already pugnacious, we have no way of knowing. Either way, the result is the same, less competition and an easier living for them and their neighboring relatives.

The behavior of the two fiddler crabs differs in a manner associated with the places in which they live. The sand fiddler returns from its feeding more and more frequently during the rising tide to check the bottom of its burrow for water level. When he finds that the rising tide has forced water to seep up into the burrow, he goes to the top of his hole, pulls sand down into the mouth of his burrow, packs it in tightly, and quietly sits just under this plug. As the water continues to rise, the sand burrow collapses; but under the plug, where the crab sits, a little bubble of air remains intact, and he waits there for the water to recede. Then he digs his way to the surface, constructs a new burrow, and is ready for the next tide.

The mud fiddler doesn't test the water level in his

burrow. There is no point in his doing so because the rising tide does not seep through the stiff, compact mud where he lives into the bottom of his burrow as it would if the burrow were in the porous sand. The hole fills only when the tide has risen far enough for the water to pour in the entrance. The mud fiddler waits until he finds himself knee-deep in water before retiring to wait out the high tide. His burrow doesn't collapse when the tide rises, and he can enter his door any time. If the sand fiddler waited too long, his home would disappear just when he needed it most, and the best he could do to protect himself would be to bury himself in the sand. Although these two species occur only a few feet apart, and are closely related, the behavior patterns of the two are thus completely different, and one will not imitate the other. We once confined some mud fiddlers in a bottomless cage in a sandy area, and they were trapped on the surface each time the tide rose.

At high tide, as we float along in the canoe, we pass over large areas of marsh that were dry when we walked on them two hours earlier. Here and there we see the prints of raccoon paws on the muddy creek banks. The creeks have now become broad, undefined waterways. The egrets move ahead of the deeper water to stand and continue their probings in the shallow areas. Some have flown off to try for food in the ponds behind the beaches.

On the surface the marsh appears calm and inactive now; but down under the water, new activity has begun. In the great reefs along the edge of the marsh, oysters living one on top of another like crowded apartment dwellers have opened their shells and are feeding. Annelids, having safely rested in their tubes during low tide, protrude their feathery filters to strain food from the water. And down in the mud, countless nematodes, earthworms, and other minute forms of life carry on continuously, shielded from the effects of tides, of day, and night, by a few inches of dense mud above them.

The ribbed mussels, which we saw earlier in clumps dotted here and there at the bases of the *Spartina* plants, began to feed even before the water completely covered them. In the course of filtering the water for food, they collect a lot of mud, which is deposited around the animals and raises the ground level; a clump of mussels can be located from a distance by the little hummock it has built. The individual mussel must adjust to this mud that fills in around him. The young ones can move easily around the clump and pick a suitable spot, but the older individuals remain in place and are gradually buried, finally sitting at the bottom of a hole kept open by their pumping.

Presently, the pull of the moon begins to drain the water down through the little creeks, into the bigger creeks, and finally out into the open ocean again; and we float with it back out of the marsh.

Production

THE SAPELO MARSHES are the great providers for countless animals. Using energy from sunlight and matter from air and water, the marsh plants manufacture food—an abundance of food that is more than enough to maintain all life within the marsh. So much food is manufactured that the creeks, tidal rivers, and even the broad estuaries are fed by it. The food consists first of the plants themselves, alive or dead; then as the plants are eaten by animals, these animals grow and serve as food for yet other animals. This growth, growth of plants and subsequent growth of animals, is the production of the marsh.

The production of food in the marsh can be studied by tracing the pathways taken by food energy as

118

it goes through the various groups of animals and plants. Using energy as the measure of production is simpler than using pounds of marsh grass and crabs, or numbers of individual organisms. This is because an energy unit, and we shall use the calorie, has the same value whether it is a calorie of grass, or worm, or fat, or sugar. It is, to be exact, the amount of heat that will raise one kilogram of water, about a quart, one degree centigrade—not quite two degrees fahrenheit. It is the same calorie that dieters and nutritionists count. Of course, a calorie of food that is made of woody plant material will not be of the same food value to a sparrow as a calorie of starchy seed. The sparrow can digest and thus use the calorie of energy in the starch, but it cannot digest the woody material. However, the plant put the same amount of energy into each package, one calorie, and if the sparrow cannot use the wood, some bacteria or fungus can.

All living things need a constant supply of energy. The constant use of energy is called the flame of life. When organisms stop using energy, they are dead. So, the analysis of production in the marsh consists of finding out how much food energy is manufactured by the plants, how much of this the plants use for themselves, how much is eaten by the various groups of plant-eating animals, and how many of these animals are eaten by the animal eaters, all quantities measured in calories.

The best place to start is in the beginning, with
the process that puts energy into the marsh in the
first place and starts the food chain. In the mud live
the tiny single-celled algae—green flagellates, dino-
flagellates, and diatoms. They sit on the surface
soaking up light energy and making themselves bigger; and when they get big enough, they make themselves into two, and these two repeat the process.
When conditions are right for the algae, a lot of
animal food gets produced. When conditions are
bad, the algae wriggle or glide into the mud for protection. Just what constitutes bad conditions isn't
clear. At times, the creek bank muds are colored
with green patches or have the uniform golden hue
associated with diatoms. In general, the algae find
the light too weak under the water in winter for
much production, and in summer they find the direct
sunlight too strong. Most of the production is under
water in summer and in air at low tide during the
winter.

Three types of single-celled algae

Here then is a ready supply of algae that can serve as food for many of the marsh animals. When the tide rises, some of the algae are stirred up into the water, and mussels and annelids that feed by filtering small particles out of the water have a chance at them. When the tide is low, the fiddler crabs work over the mud. The algae-eating nematodes that live in the surface layers of the mud eat all of the time, both high and low tides. The algae production adds up to some 1800 calories per square meter of marsh per year.

This algal production compares favorably with the production found in many lakes and parts of the oceans; but the algae account for only one-fifth of the total plant production of the marsh. The *Spartina* produces the remainder. Its production energy is contained in the tissue of the plants.

Since plants are very much "living things," they use some of the energy they capture from the sun for their own life processes. The marsh grass actually uses a larger proportion of the sun's energy it captures for its own life processes than do most plants. Ordinarily, ecologists figure that plants use about one-tenth of the energy they fix from light for their own processes and build the other nine-tenths into their tissues. The marsh grass sometimes uses up more than half of the energy it fixes. This is due to the difficulty of living half-in and half-out of water.

Other grasses are watered with fresh water, and their essential processes are geared to a large, readily available supply of fresh water. *Spartina* grows in water containing three percent salt. But since it is essentially a land plant, with land plant ancestors, and although it has adapted to living in this salty water, it must get rid of the salt in order to use the water. That is a process that takes energy, as those engineers working on processes to make fresh water from the sea know only too well.

Spartina does very well in its difficult situation, however. In spite of the energy it must expend in desalting its water, it still manages to do a lot of growing; the total growth in a year is very impressive. When the *Spartina* production is added to the algal production, the marsh is seen to rank just below the most productive areas, like sugar cane fields, and far above the average for the world.

This success story is due to the tides. For although it is the tide that brings the salty water that gives *Spartina* its problems, the tide also stirs the marsh twice a day. The waters mix the mud and rinse the plants, and in doing so remove waste products and bring fresh nutrients to aid plant growth. The plants of the marsh actually capture, and either use or store, about six percent of the sunlight that falls onto the area per year.

While the fiddler crabs are eating the algae, the marsh insects are eating the marsh grass. Grasshoppers eat the tender leaves during the long summer; and plant hoppers suck the juices all year round, but are more abundant in the winter. Still, the insects eat only a small part of the grass, and most of it is left to die. Then bacteria from the marsh mud and the seawater begin to decompose the dead leaves and stems so that they break up into smaller and smaller pieces. Finally the grass is transformed into what is known as detritus, tiny particles that look like bits of mud or dirt but are actually remains of plants and the bacteria that have been growing on them.

Detritus is the most important foodstuff of the marsh. It is constantly produced as the old leaves on the plants die. The great mass of plant material that dies after seeds are formed in autumn decomposes slowly during the winter cold. All in all, the supply of detritus is fairly constant throughout the year.

Most of the algae-eating animals also eat detritus and probably get most of their food from the latter source. Some of the nematodes are exceptions, since they are so small that they can pick out individual particles to feed on and some pick algae in preference to detritus.

The plant eaters are food for the animal-eating marsh dwellers: raccoons, marsh hens, mud crabs, and herons, which together eat about forty percent of the total annual production of fiddler crabs and snails. After all the consumption of the marsh is added together, and after the number of individuals necessary to maintain the population and serve as the basis for the next year's production are added up, nearly half of the marsh production remains. It might seem that the salt marsh has a problem of surplus food that makes the problem of our North American farmer look ridiculous. This is not the case, however, and the reason lies again in the tides and the water currents they produce.

One-half of the marsh production is stirred up by the water and carried away before the marsh animals ever have a chance to feed on it. The tides, by bringing nutrients and removing wastes from the plants, make a very high marsh production possible; but at the same time, they rob the marsh consumers of about half of this production, putting everything back on a level with less favored spots in nature.

Because of the strong stirring action of the tides, the waters of the tidal rivers and sounds are very muddy. Light can penetrate only a short distance into their depths. So the algae, the phytoplankton, that normally live floating in coastal waters and are the ultimate source of food for the animals, are scarce and produce very little. Thus, the phytoplankton-eating animals must find their food elsewhere. This is where the marsh helps out. The myriad shrimp, baby menhaden and blue fish, edible blue crabs, bottom dwelling annelids, oysters—all depend on the marsh. The marsh plants and animals produce food. The tides carry part of this food away from the marsh and to the animals that live on and in the bottom. And the tides even carry swimming animals into the marsh to the food.

The marsh today continues to live with its overlapping and interacting cycles of seasons, days, and tides, in the same way it has for hundreds of thousands of years. It has been used to a limited extent for cattle food. Cattle graze the portions nearest land when their winter pastures are brown. Men sometimes cut the grass to carry away as cattle feed. But this seems to be about the limit of use to which man has put the marsh. There have been various suggestions that legislation be passed or enterprises established to convert the "waste land" of the marshes to useful purposes. It might be diked and used for farming. This looks like a splendid idea to those accustomed to thinking of crops in farming terms; but any seasonal crop that man might grow on diked marsh could not compare with the production of those miles and miles of grass, the tips of the plants all reaching about the same level, as though they had all been trimmed by a giant barber. It should not be forgotten, too, that man's crops would have to be fertilized at considerable cost. The sea fertilizes the marshes. And most important, diked marshes would not provide food for the animals that maintain the shrimp, crab, and fish industries. The marsh is of inestimable value to man just as it is.

PART IV

The Beach

Beach Drift

THE LAND, the marsh, the beach,—only the last is normally thought of as the edge of the sea; but all are involved in dividing the water from the mainland. The land of the island is the part least obviously involved in this division. However, a comparison of the island with the mainland shows how important the sea is, for it helps maintain live oaks, Spanish moss, palms, and other semitropical plants dependent on warmth and humidity for life. These plants exist much farther north on the sea islands than they do on the mainland, which is separated from the warming sea by a barrier of islands from North Carolina to Florida.

The beach, however, is the obvious, literal edge

of the sea, the last thing one can walk on until coming to Bermuda or the Canaries, or Morocco far to
the east—so far to the east that the sun rising out
of the sea at seven on a winter's day on Sapelo is at
that moment at its noon height in the sky over the
land at the other edge of the sea.

The boundary of the beach changes a little with
each tide, and is modified in a cycle as the moon progresses from crescent to full and back again. Occasionally, a high wind will drive sea water up over
its normal height into the low area behind the beach
ridge facing the sea, but generally the water lies in
front of this ridge of sand built up by tides, waves,
and wind. On the front of the ridge a few succulents
and sea oats grow. Farther back from the sea, myrtle
and pines begin to appear. The pines are found in
the depressions between the dunes, and the ridge of
dunes is often capped by stunted, wind-pruned oaks.
These give way finally on the land side of the dunes
to larger oaks, pines, hollies, and other land plants.

The tides and surf continuously carry plants and
animals, torn from their homes in the water, and
dump them at high-tide line with an occasional pack-
ing case, old rubber boot, orange rind, and net float.
We would often stroll along the line of beach drift,
looking for treasures.

Here and there along the high-tide line, especially
after strong onshore winds had blown the surface
water of the gulfstream toward land, we would find

the delicate, transparent, blue floats of the Portu-
guese man-of-war. The wispy tentacles would already be nearly dried; but even then, out of their
milieu and under a hostile sun, they still packed a
poisonous wallop; so we were wary as we combed
the debris. After such a wind, the debris would consist largely of *Sargassum;* its brownish leaves and
hundreds of little amber floats would tumble over
the sand in a mass, like so many wigs done up in
buns, held in place by red and yellow branching
coral for combs, which were bejeweled with pink
snails. Stalks of old *Spartina* plants stuck through
the buns like so many hat pins. Hundreds of amphipods and beetles live in these masses of vegetation,
feeding on the rapidly decaying material.

On other days, the high-tide line wears entirely
different clothes: perhaps a skirt of sea cucumbers,
some of which have coughed up their stomachs in
the trauma of being dumped high and dry; or a
petticoat of jellyfish, which dry on the sand to trans-
lucent discs measuring from one inch to a foot in
diameter.

There are always the shells: fragile, white angel
wings; disc shells; blood arks; channeled ducks, so
delicate it is a wonder that they are carried up unbroken; the great heart cockles; lettered olive shells;
common razor clams and stout razor clams; moon
shells; whelks; the tiny cochina or butterfly shells;
baby's ears; and on and on.

A few types of sponges come ashore regularly. The large, hollow, dung-colored loggerhead sponge houses a variety of animals in its caverns. One is a tiny crab with a bright red body covered with a dense forest of black tipped spines and hairs. By breaking through the walls of the caverns one can also find baby sea stars and brilliantly colored, orange serpent stars, only a quarter-inch across.

Occasionally, a larger animal is caught in the shallows, tries to flounder its way back to deeper water, and fails. On a falling tide it comes in too far on a big wave. When the wave recedes, it is left stranded, unable to get back into the water until the subsequent high tide, and by that time it is dead. Two pigmy sperm whales were stranded thus on the beach. One was small, only one hundred thirty-five pounds, and the other was a mature female in nursing condition. We suspected that the young one became stranded and the double tragedy resulted when the mother tried to aid her young. They were stranded during the night; for when we found them the next midmorning, they were already dead and half buried in the sand. A group of vultures circling overhead was beginning to settle for a meal.

The pigmy sperm whale is seen very rarely, although it has turned up in widely separated places and may be more common than records indicate. It is rare enough as far as zoologists are concerned. Our young whale was embalmed and shipped to the

National Museum for study. We dissected the female at the Marine Institute and found that the stomach was full of squid beaks, indicating that these swift animals, which can dart at incredible speed, must be the whale's chief food.

Besides the plants and animals that wash up out of the sea accidentally, there are certain animals that come out of the water intentionally to lay their eggs. The loggerhead turtle does this, as does the horse-shoe crab, *Limulus,* not a crab at all but an arachnid, a distant relative of scorpions and spiders.

Limulus, a relic of past ages, probably lived in the area when the sands that would later form the backbone of Sapelo were only beginning to accumu-late. *Limulus* then fed on the animals living on the later-to-be-Sapelo. Now they use the island's beaches as places to hatch their eggs. The crabs crawl ashore at night in pairs, the female, who may be a foot wide, ahead, dragging the smaller male, who hangs on to her with his specially shaped front feet. The female lays her eggs in the sand at about high-spring-tide level, and the male spreads sperm over them as they are laid, much as frogs do when they are laying

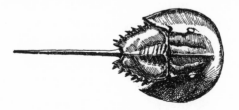

LIMULUS, *or Horseshoe Crab*

their eggs in ponds. The two then separate, each
trying to find his own way back to the water. Often
they fail and are found next day, killed by the heat
of the sun, at the end of a trail that twists over the
beach in what seems a random fashion. *Limulus*
doesn't have much in the way of equipment with
which to find its way back to the sea, a distance of
perhaps five to twenty yards. Its two primitive eyes
are immovably imbedded in its horseshoe-shaped
shell and can do little more than distinguish light
from dark, although they are also sensitive to the
plane of polarization of light.

The loggerhead turtle is considerably more impressive as she visits the shore. First, her size is impressive. A horseshoe crab is of considerable size in
comparison with most animals found on the beach,
but the loggerhead may be three to four feet long
and weigh more than three hundred pounds. Only
the female turtle comes on the beach. Since mating
takes place somewhere at sea, the males are spared
the hazardous and difficult terrestrial journey. Prog-
ress on land is laborious for the large turtles; they
are better adapted for swimming in a medium that
supports their bulk. On land the female moves by
lifting itself up and forward with its flippers; but it
cannot get the flippers in under its body to support
its weight. The turtle's effort resembles the effort
you would make if you tried to row a boat through
sand. Still, the loggerheads have an easier time than

the very rare leather-backed turtles, which may weigh nearly a ton.

The loggerhead turtles do not come ashore in equal numbers on the various islands along the coast, but seem to favor Jekyll Island, second south of Sapelo. Why this should be is not at all clear since Jekyll, to the human eye at least, does not seem to have anything that makes it especially suitable either for turtle nests or for helping the females get to and from the beach. But go the turtles do, in numbers that exceed those that visit all of the other Georgia islands added together.

Studies of turtle crawls were made from an airplane that flew over the beaches along the coast. It is easy enough to spot a turtle crawl, which is seen as a trail leading from the water to high-tide mark and back. Think of yourself in the boat again, rowing through the sand, and visualize the groove dug by the body of the boat with oar marks at the sides, which correspond to flipper marks. The only difference between the boat track and the turtle crawl is caused by the turtle's tail. On the way up the beach, before the turtle is exhausted, her tail makes short, sharp marks in the middle of the track. The tail mark on the return trip is one long furrow. The egg-laying process so exhausts the turtle that she cannot even hold her tail up at the end of it.

Watching a turtle is a fascinating experience, although it means missing a good part of a night's

sleep to see it through to the end. One must not dis-
turb the animal as she comes out of the water, or
she will return to the sea without laying; but once
she has gone to the effort of climbing the beach and
beginning the nest, one can look at her quite closely.
Even a few flashbulbs do not upset her efforts.

Turtles come up onto the beach only when the
high tides of early summer come in the first half of
the night. Usually, we should say rather than only.
We did see one animal laying eggs under a hot midday sun on St. Simons Island.

The high tide helps the heavy turtle get up the
beach and also indicates a minimum level beyond
which the eggs must be placed if the developing
young are not to be drowned by subsequent high
tides. On nights when there was a full moon com-
bined with such a tide, we often watched for turtles.
It was easy to imagine ourselves back in the age of
reptiles. There was no noise on the beach, except for
the scurrying of ghost crabs in their nightly quest
for food and the gentle wash of the surf just ahead
of where we were sitting, high on the beach. The
turtle, when she came, let the waves carry her as far
as possible up the beach. She rested high and dry for
some time, looking toward land. Finally the lumber-
some shape, only now really recognizable as a turtle,
began the slow crawl up beyond the high-tide mark
and along the beach until she found a spot that
suited her.

Then she began the real work. She dug alternately with her hind flippers, which are flexible and capable of quite precise movements. She cupped the end of the flipper and scooped up a flipperful of sand, raised it and placed it carefully to one side of the hole. She then rested that flipper while she dug with the alternate one. Just before putting the first back for more sand, she pushed vigorously to the side, scattering the sand dug out previously, in a sort of mechanical ballet. The hole that resulted was circular, somewhat larger at the bottom than at the top, and about two feet deep.

When the hole was finished, the turtle extended her cloaca down into it and eggs, about the size and shape of golf balls, began dropping to the bottom.

Loggerhead Turtle

For each egg laid, she moved her front flippers slightly. About midway through the operation, the eggs began coming in twos and threes and the intervals of rest between eggs increased from ten to twenty or thirty seconds. All the while she was laying, heavy tears ran down her face, the production of her tear gland, which in turtles, as in other reptiles and sea birds, secretes the extra salt the animals acquire by drinking sea water rather than fresh water. This weeping goes on continuously, but is not noticeable except when the female is ashore making her nest.

She finally finished, after laying one hundred and seven eggs, and then began flipping sand with her hind flippers. Occasionally she picked up a flipperful and placed it in some exact spot, only to scatter it moments later. During this scattering of sand, she moved about slightly until she had disturbed the sand over an area of several square yards. Then she started back toward the water. It was difficult to find the exact location of the nest even though we knew it was within the disturbed area and had actually seen it dug.

During the egg-laying, the turtle got visibly more and more tired, and was near total exhaustion as she started down toward the water. She became confused and dragged herself a wasted distance parallel with the water, making the trip twice as long as necessary. Even if she had gone straight to the water, the journey back would have been longer than the one up the beach, since the tide goes out while nesting takes place. The turtle rested frequently on the trip down. Only when she felt the surf lift her a little did she seem to exert a last bit of energy to make an effort to stay with the wave as it receded from the sand. On the third large wave, which lifted her bulk free, she was able to float out to sea and rest.

The eggs lie in the sand while the embryonic turtles grow slowly inside. It takes about two months for the young turtles to grow to hatching size. A number of hazards endanger the turtles during this period. Any animal that can find the nest, including man, looks upon it as a bonanza. Ghost crabs are frequent pillagers, building their burrows so that they can go down into the egg store, which assures them of a well-stocked larder for the rest of the season.

Another hazard, depending on when the eggs are laid, is rain. In the autumn when the heavy rains fall, the beach and dunes become completely water-logged. The rain can't sink very far into the sand because of the salt water that has seeped in from the sea. So it collects in the upper reaches of the beach and drains slowly through the sand. Turtle nests are flooded during any such period, and any unhatched eggs drown. Since there is such a long period between laying and hatching, any nests made after the first of July run a grave risk of being destroyed.

If the eggs are not eaten or drowned out and the young succeed in hatching, they meet other hazards. The little turtles find themselves deep in sand and must dig their way to the surface. We have dug into nests after the young have emerged, and have found some that succeeded in getting out of their shells but failed to reach the surface, perhaps because of exhaustion. Once they reach the surface, the young turtles instinctively head for the sea, running the gamut of birds and animals who are most happy to dine on tender turtle. If they make the journey to the sea, they head immediately for deep water and begin to contend with fish enemies. Years later, any females that have successfully overcome or evaded all these hazards return to a sand beach somewhere to start the cycle again.

Zoning

THERE ARE A large number of animals that make the beach—like the marsh, a transition between water and land—their principal or only home. The marsh dwellers, however, are adapted to a watery land, characterized by mud, quiet water, and the stabilizing influence of marsh grass; and those choosing the beach as a home find it a place of unstable sands and surging waters. They are adapted for living in an area that extends from just above the level of the highest tides to the level of the lowest exposed parts of the beach at low tide, more under water than above. This area is one continuous realm, with animals overlapping in their habits and requirements, but it may be divided into three zones. The animals

141

in the zone farthest from the water live mostly in air or burrow in the sand. Some are typical land types that have grown accustomed to the beach. Most of the animals in the second zone are marine forms that have made the adjustment to being out of water part of the time. The lowest reaches of the beach, or what we will call the third zone, house only animals derived from the sea, which hide in wet sand or water-filled burrows during the short low tide they experience.

The beach, in many ways, is an extremely unpleasant place in which to live. The soil is unstable, shifted by each tide, thoroughly churned by storms; anything projecting from the surface is given a sand blasting with each high wind. The landward edge is a saline desert because occasional spring and storm tides bring sea water that evaporates, leaving its salts behind. The plant life, therefore, consists of only a few succulents just above high water. Paradoxically, fresh water is also a danger; for a marine animal may find the rain soaking its precious salts away if it ventures too far up the beach. Without plants to break the winter gales or provide shade from summer sun, the beach can also be very cold and very hot. But animals live there in large numbers.

There are some advantages. Food is concentrated there by the water. Flotsam collects along the high-water mark. Waves furnish a constant supply of fresh sea water for filter feeders, without any effort on their part. Finally, some animals may be able to make a living in a marginal environment who couldn't stand the competition in more favored places.

Rove beetles and tiger beetles live in the upper beach area. The rove beetles, slender and about one-half inch long, are somewhat smaller than the tiger beetles. The burrows the rove beetles make as adults have holes about half the size of those made by the tiger beetle larvae. Their small size is compensated for by super abundance. There may be a rove beetle burrow every inch in every direction in some areas. What these beetles feed on hasn't been determined, but it is safe to assume they are not predators but live on the general beach drift. Spiders also inhabit the upper beach. Like all spiders, they are predators and concentrate on the rafts of beach drift which harbor small edible animals.

These are typical land animals, but also living on the upper beach are amphipods and crabs, whose ancestors came from the sea. Pale-yellow-to-white ghost crabs, about six inches across including legs, are rarely seen by day, but are at large in great numbers at night. The pale body of the crab, rushing about in the moonlight, attracts the eye; but as it comes to a sudden crouching stop, it seems to disappear.

These crabs live in the dunes, sometimes a hundred yards from the water. There they dig a deep burrow in the damp sand where they wait through the hot day. At night they emerge and look for stranded animals, turtle eggs—in short, anything edible. Often the end of the trail for a horseshoe crab, in to lay eggs and unable to regain the sea, is the stomach of a ghost crab, who is very likely to dig his burrow near the stricken animal when caught in his feasting by the approaching day.

Ghost crabs must go to the surf occasionally to keep their gill cavity moist and to lay eggs, which then become part of the ocean's plankton. But the crabs have made considerable advance toward land life; for their gills are very small and the gill cavity has developed into a sort of lung for air breathing.

The beach amphipods, or beach hoppers, rather distant relatives of the crabs, are not so much adapted to land as adapted to avoiding the effects of land-living while staying near high-tide level. There are several species adapted for slightly different habitats, but all are robust amphipods about an inch long as adults. Ghost crabs are occasionally seen during the day; beach hoppers, never. Ghost crabs move one hundred yards or more from high tide mark; beach hoppers not more than a few yards. The amphipods spend the day in a burrow plugged with sand against enemies and the drying effects of air and heat. They emerge at night to feed on plant and animal material, then before dawn dig a new hole in which to hide.

At the second zone, around mean-water level, where there is water as often as not, a number of typically marine animals live. Several of these animals dig burrows at least three to four feet deep in the sand.

Two species of ghost shrimp, *Callianassa major* and *C. stimpsoni,* are common. *C. major* grows to be up to six inches long and makes holes one-half inch across; *C. stimpsoni* is only one to one and a half inches long, makes small holes, and lives in somewhat muddy sands. During low tide, both stay deep in the sand where the salinity and sand are stable and there is no danger of drying. They come to the surface to feed on bacteria, algae, and detritus, which they clean off sand grains when the tide is high. Because there is so little oxygen deep in the sand, they must circulate the water in their burrows to keep them aerated. This they do with specially

CALIANASSA MAJOR *and* C. STIMPSONI

adapted paddle-shaped legs, which lie along their abdomens. This has the effect of oxidizing the iron in the beach sands, so each hole is surrounded by a tube of orange-colored sand cemented together with rust, distinct from the surrounding gray of deeper sand layers.

A number of segmented worms, annelids, inhabit the middle beach zone. Closest to the upper beach is the long, reddish, *Onuphis,* which is about as big around as a lead pencil and eighteen inches long. It builds and lives in a tube lined with a thin parchment-like substance. Sometimes great numbers of these collapsed linings accumulate on the beach, washed out of the sand with the changes in beach contour.

The worms, like the ghost shrimp, feed on detritus, bacteria, and algae, and must clean large quantities of sand in order to collect enough food to live. They do not swallow the sand, but separate it out before they swallow. This results in the production of quantities of fecal pellets, tiny lumps of mud including no sand, from which the food was digested in the animal's gut. Most of the animals void their fecal pellets on the surface of the sand. Since the pellets are of different density from sand, they tend to accumulate in layers on the beach. So many of the pellets are produced that if one sieves some sand gently through a mesh just large enough for most of the sand to be washed through, most of the remaining, unsieved material will be made up of these tiny pellets. They will be of slightly different sizes and shapes, each variety characteristic of the species that produced it.

Some animals filter the water for their food. They move up and down the beach with the surf. They are always found at the level where the backwash of the surf provides a current full of the fresh supplies of food that come in with each new wave. Three particularly numerous representatives are: mole crabs, coquina clams, and *Haustorius*. The mole crab, a peculiar crustacean adapted for digging backward in sand until just his mouthparts and antennae, by which he feeds, show, is the largest. The coquina clams, with colorful vari-hued striped shells, are the next largest; and *Haustorius,* a one-quarter-inch amphipod like the beach flea, is smallest.

Their skill in burying themselves allows these three creatures to live successfully on the beach. Many beach dwellers can bury themselves, but it takes them several minutes to disappear even when not bothered by the gentle surf. Consequently, they

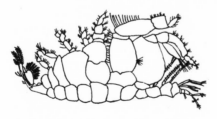

HAUSTORIUS *in swimming position*

don't come out of the protective sand except by accident. However, *Haustorius* can bury himself in one second, and the coquina and mole crabs require only a few seconds. These swift diggers can afford to live close to the surface where the greatest abundance of food may be found. They pay for this advantage by being the animals most often caught by the shore birds as they probe the sand just ahead of the waves. On Sapelo beaches, where coquina and mole crabs are relatively rare, most shore bird food consists of the tiny *Haustorius*.

Most of the species encountered on the upper reaches of the beach are herbivorous or omnivorous. They eat algae and detritus from the water or dead animal and plant material that washes ashore. Lower down there is more food simply because the water is there more of the time. With more animals, there is a chance for carnivorous animals to make a living; and so it is close to low-water level in the third zone that most of the beach animal-eating animals are found. They consist mostly of snails and worms that can burrow in the sand and avoid wave turbulence.

Pink ribbon worms two feet long occur and feed with the aid of a proboscis, which can be shot out to entangle their soft bodied prey and retracted when not in use. Carnivorous polychaete annelids live in this zone, too. One is *Glycera*, which looks something like an earthworm, but has a thick, short proboscis armed with four sharp hooks on which it impales its food.

Snails are important predators. Most of the various bivalved molluscs whose shells wash up on the beach bear a little beveled hole near the hinge of one shell, the mark of the moon shell snail. The hole is made in the living bivalve by the snail with its radula, a flexible band set with teeth, which snails use as rasps or files. First it rasps the hole in the shell; then uses the same tool to eat the meal so exposed. The voracious snail, whose enormous foot when expanded almost covers the shell, can sometimes be found under the sand at the end of a little ridge similar to the one a mole makes on land. Each ridge is the product of the ceaseless search for food.

Whelks also eat clams, and sometimes use what seems to be a remarkably clever way to open the shells of their victims. The whelk grasps the clam with his foot and wedges the edge of his own shell against the place where the clam's shells meet. By wedging and twisting, he proceeds to chip away the shells, including his own, which is later repaired, until there is an opening big enough for him to insert his proboscis and rasp out the tender clam flesh.

In *Terebra,* another animal living in the third beach zone, the teeth on the radula are little hypodermic needles equipped with poison sacs. *Terebra* are related to the tropical cone shells, the rattlesnakes of the sea. Sapelo's *Terebra* is only one to two inches long and harmless to humans, but its tropical cousin can inflict painful wounds that sometimes cause death.

Terebra Shell

There are more animals, as well as more kinds of animals, as one goes lower and lower on the beach and the exposure to drying air and heat is lessened. So, reasonably enough, certain creatures are found in tidepools and beach sloughs that don't occur on the beach at all otherwise. Hermit crabs and sea pansies are the most noticeable.

The hermit is a crab-like animal that uses a snail shell to protect his soft body, trading it for progressively larger ones as he grows. The sea pansies are coelenterates, relatives of sea anemones and jellyfish. Each pansy is composed of a number of individuals, some of which have tentacles like coral animals or anemones. These feed the group, and those, without tentacles regulate the water pressure and the rigidity of the colony; changes in water pressure are the means by which the stem is thrust down into the sand to anchor the pansy.

Besides the changes in the beach life from high to low tide, there are also changes caused by the nature of the beach. Some places, more protected from the surf, are muddier than others where the waves wash all the fine particles away. Animals are more abundant in the muddier places because there is more food. An appreciable part of the mud is organic detritus from the rivers and salt marshes. Also in the quieter areas there is less disturbance from wave action.

At the lowest levels in these muddy spots, the surface may be covered with little holes and mounds, indicating the life buried beneath: tiny holes where the arms of a burrowing brittle star reach the surface; medium-sized holes housing vicious sickle-armed mantis shrimp; and holes one inch across and three or more feet deep, at the bottom of which lie the delicate-shelled, six-inch angel wings. Mounds and coiled castings indicate a group of our extremely primitive and distant relatives, the acorn worms. Their long rear sections are so delicate that two special techniques for digging them out are used. One is to approach carefully, reach out and set the spade very gently on the surface, and then suddenly jump on it forcing it into the sand and turning up a spadeful with the same motion. This often exposes a wary animal before he can flee. The other technique is to dig a large deep hole and then carefully scrape away the sides, exposing the delicate worms with a minimum of disturbance. The latter method is not recommended for any but the most interested; for usually by the time the hole is dug, the neighboring animals have left.

By either method one usually does a great deal of work for little reward. Once, however, we were amply rewarded for our effort. In a last quick spade thrust, made just before the tide covered a sand flat, we turned up a bright purple-red proboscis worm whose plump nine-inch body was studded with neat, wartlike projections and terminated in a three-inch yellow proboscis. From the same burrow, we drew a male and female of a bristly pea-crab species, whose home until then was unknown.

The *Noctiluca,* a tiny, one-celled dinoflagellate, true to its name, lights up the night when disturbed with a pinpoint of light. On occasion, quantities of *Noctiluca* are washed ashore by the waves and left on the sands, where, at night, they live for some time. On moonless nights we've walked along the beach just above the waves and looked back to see our footprints etched by thousands of points of light, fading out a few yards behind us to leave the beach to the sand, wind, sea, and animals as it had been before we trespassed.

Birds of the Shore

DAWN ON SAPELO'S BEACHES often sees a low bank of clouds lying far offshore. The horizon bursts into flame as the sun burns its way through, scattering a shimmer of gleaming color on rippling waters near shore. Shore birds, quietly flying by on early morning business, are silhouetted against the brilliance. Above the sound of the soft, slowly-dying land breeze, a gentle surf raises its lovely, delicate voice.

The sun climbs quickly, warming the air and reversing the breeze so that it blows from the sea. The shore birds stand out in detail as the blaze upon the water diminishes.

Some shore birds are present throughout the year. Most noticeable are the pelicans, their off-white

153

heads bent over against their brown breasts so that
their extravagant bills will have solid support. They
sit quietly on the sands, as dignified as monks in
groups. And they fly beautifully, sailing in long files
just over the crest of the waves, with their long
wing-tip feathers almost touching the water. Only in
their food gathering do they lose their dignity. They
fly some twenty to thirty feet up looking for fish, then
dive down after them. Just before they reach the
water, they stretch their wings out behind; then they
hit with a splash that must frighten the fish for yards
in every direction, but doesn't save the one their
beak closes upon.

With the pelicans on the summer beach are black-
headed laughing gulls, large-crested terns, orange-
beaked royal terns, and a few red-beaked Caspian
terns and yellow-tipped black-beaked Sandwich
terns. There are groups of black skimmers, amaz-
ingly graceful birds, which, as they fly along with
their lower bills cutting the water, seem to be flapping their wings in an effort to stay down near the
water rather than up out of the water. In the early
morning and evening, they can be seen fishing the
shallow water off the beach and in the sloughs, occasionally flipping a fish into their mouths.

We visited nesting colonies of skimmers on one of
the tiny sand islands near Sapelo, Little Egg Island,
and found the eggs and young lying in shallow depressions in the sand. Both eggs and young, but

especially the young, blend in with the color of the sand so well that they are in great danger of being stepped on by large visitors like us.

Some of the other shore birds nest on Sapelo. There were two pair of oyster-catchers, large black-and-white birds with striking red bills, nesting on the beach; but only one year in four did we see young that survived to flying size.

This species is very territorial. A pair stakes out a mile or so of beach and demands that the territory be free of violation before it will nest. With the development of so many Atlantic coast beaches as resorts, the oyster-catchers will decline in number. Fortunately, they are long-lived birds. We saw forty oyster-catchers on the beach during one fall migration, which represented nearly the entire population of the Georgia coast. They were all adults. No young were produced that year by any of the twenty families.

Three other birds, willet, least tern, and Wilson's plover, nest on the beach also. The latter two nest in a rather loose colony, the nests of all the pairs fairly close together. They make a shallow depres-

Oyster-catcher

sion in which they lay their eggs. Their young, as is
the case with all shore birds, are active as soon as
they have hatched and are able to protect themselves
to a limited extent. They also blend in with their
sandy surroundings so that, by lying motionless, they
are often missed by predators who hunt by sight,
such as hawks and gulls. The small shore birds nesting on Sapelo must also survive the predations of
raccoons, which hunt by scent, and rattlesnakes,
which hunt by detecting the warmth of warmblooded animals. These birds are obviously best
adapted for nesting on tiny sand-bar islands too
small to support resident predators.

The beach-living willets are quite solitary in their
nesting habits. Some willets nest in parts of the
marsh where they build their nests quite close together. On the beach, however, the nests are widely
separated. There are willets on the beach at all times
of year. The summer willets, which nest on Sapelo,
spend the winter in Venezuela and neighboring coun-
tries; the birds that winter on Sapelo spend their
summers in northwestern United States and south-
western Canada. The two look very much alike, but
they are, in fact, separate subspecies.

Toward the end of the nesting season, the hordes
of shore birds that nest in the tundra far to the north
begin to come south, and the beach is crowded
with thousands of individuals. Certain species pre-
dominate: red-breasted knots, black-bellied plovers,

red-backed sandpipers or dunlins, hunchbacked sanderlings, western and semi-palmated sandpipers. At times any one of these species may be represented by thousands of individuals. Flocks will be at the water's edge for miles along the beach, following the waves in and out, probing the sand for the little crustacea and worms on which they feed. At the southern-most tip of the beach is a large, open, flat area, much of it sand, some covered with short grasses. Here the larger shore birds are seen in great numbers. Hudsonian curlews sit with their long down-curving bills among the marbled godwits with their long up-curving bills.

In the winter, the most prominent birds are the gulls and terns. Forster's terns occur in flocks of

Black-bellied Plover and Dunlin

hundreds. Herring, ring-billed, and Bonaparte's gulls are in evidence all along the beach. The herring gulls have different plumages in the first, second, and subsequent years. The adult dress of white and gray follows a mottled brown in the young birds. The first herring gulls to come south from the nesting grounds are the youngest, birds that have been raised only that year and have never traveled the flyway before. The yearlings arrive next, and only in the middle of winter do the adults, who have traveled the route a number of times, finally arrive.

Other birds are present in large numbers in winter, but they are not so evident along the beaches because they spend much of their time swimming. Only occasionally do they come onto the sand. The cormorants come ashore more frequently than any of the thousands of lesser scaup ducks, which spend most of their time in sounds and in the ocean just offshore.

Birds, other than the aquatic ones, also frequent the beach because of the ready supply of food that washes up out of the ocean. Especially evident are the fish crows and vultures. The fish crow is much like his slightly larger cousin, the crow, in habit and appearance, but is more inclined to come to the shore to feed.

Vultures depend to a large extent on the shore to supply them with food. Land animals die in numbers large enough to keep many vultures supplied with food, but a much larger vulture population is able to exist because of the food thrown up on the beach. There are always a few vultures circling on the sea breezes that come in over the beach and rise as they come to the dunes. When one of these watchers sees a fish, or better, a porpoise or sea turtle, wash up at the high-tide mark, it promptly spirals down to land near the food. Others see the first and follow, and in a short time there may be six to twelve birds sitting around the animal, waiting for something to happen. If the animal doesn't move, they gradually walk or hop closer until they finally begin to feed.

Once we walked around a dune and saw thirteen black vultures clustered about a stranded sting ray. They were handsome with their matt-black heads contrasting with the glossy feathers on their backs. They jumped away as we came near and twelve flew off to neighboring dunes to wait for us to leave. The thirteenth began to circle slowly overhead and, as we watched, rose steadily until it reached a great height. We walked on, and the birds returned to their feast, the one descending rapidly. This was unfortunate for the thirteen diners, for the descent attracted the attention of other vultures. They came, one at a time, sailing over at such great heights that they were difficult to see. On reaching a point directly over the group on the ground, they spiralled rapidly down. By the time we left twenty minutes later, there were eight new arrivals to share the food.

There are two species of vulture, the red-headed turkey vulture and the black vulture. The turkey is the better flier and can sail with motionless wings on updrafts too slight to support the black vulture. The black is definitely boss at the table, however, and sometimes one will take charge, chasing all others away, especially if the food item is small. Even less favored black vultures manage to sneak mouthfuls of a large food item, but the turkey vultures are always at the bottom of the pecking order.

It is reported that in some places in the South, the supply of dead animals is getting so small, because of modern farming techniques, that the vultures have occasionally taken to killing their food. On the uninhabited beaches, there is still a plentiful supply of already dead food; but with the continued development of Atlantic beaches with summer cottages, hotels, and all the other appurtenances of modern civilization, perhaps times will get hard for the maritime vultures as well.

Finish

THIS IS WHAT we saw of Sapelo in our four-year tenure. We came, basked ourselves in the pleasures of one of the most beautiful spots on earth, as yet unpolluted by an abundant population strewing its waste over the landscape.

We were able to walk four miles of white sand beach with only a few unconcerned animals for company. We were able to penetrate dense forests and attempt to unearth their secrets. We were able to watch the succession of plants after forest fires, to see the destructive force of storms, to watch the soil of the island grow and diminish, to see the ebb and flow of gentle tides, and the crash of surging waters. We were able to see the cyclic wealth of production in the salt marshes, to observe countless birds on the wing, and to watch other animals playing in and over the land.

And then we left; but always remembering the joy of seeing new facets of nature each day to put in our Portrait of Sapelo Island; always impressed when we remember by the variety and the bounty of life we have witnessed.

161

Afterword

WE LIVED on Sapelo Island from 1955 through 1959 and after a twenty-year hiatus we returned.

As we drove along the highway from Savannah through Darien, then backtracked a short distance to Meridian, and finally turned onto the well-remembered road leading to the dock where the Sapelo boat tied up on the mainland, we were struck with just how little the area had changed.

Shrimp boats, put up now for the season, settled in the shallow waters of low tide. Tall, straight-trunked pine trees, live oak, holly, and palmetto lined the way, looking much as they had when we first saw them. However, we had changed considerably. White hairs had grown in among the dark, we

163

didn't move as quickly as we had, and whereas the area seemed vast before, after our constrained city living, it had shrunk a little to our much more world-traveled eyes.

A modest amount of litter, carried on high tides, lay about on the marsh surface to either side of the causeway. It had as well twenty years previously. Perhaps it now contained a greater amount of plastic, but plastic was already plentiful in the early fifties. In the natural landscape it was very much "business as usual." The birds still called out as they circled overhead and small waves still lapped at the banks of the waterway. We wondered how many generations of gulls and herons had come and gone. Still, their descendants flew over the waterways looking and sounding the same.

Even the man-made landscape had changed little. A few houses had been built along the way. The parking area at the dock had been enlarged for commuters, and more cars parked nose-in every day. Gone now was the *Kit Jones*, the old converted tugboat that had been our ferry—our lifeline to the mainland—and coming to meet us was a fast, low-slung commuter boat. Gone, too, was Chief Olsen, the old redfaced sailor who had piloted the *Kit Jones*. The *Kit* had often been joined by two dolphins who alternately played around her bow and then raced ahead to show her the way through the Doboy Sound.

Perhaps the biggest natural change had been the

disappearance of Spanish moss due to disease. The heavy festoons which used to hang nearly to the ground had dwindled to meager hanks. There is still much around, but the healthy growth prevalent through our years on Sapelo was gone. Whether or not the moss will recover to its former abundance is still an open question.

The trip through Doboy Sound and up into the Duplin River took a quarter of the time it had in the *Kit Jones* or for that matter in the *Caretta*, a small inboard on which we risked our life regularly during rough weather and through the strong currents flowing into Doboy Sound. When the boat docked, several children got off, on their way to a schoolbus parked nearby, and several people boarded who lived on the mainland and worked on Sapelo for the Fish and Game Division of the Department of Natural Resources. On docking we found that the pilings still carried their load of oysters and barnacles, but now instead of climbing a ladder, hanging six feet over the side of the dock for use at low tide, we walked up a floating ramp.

As we drove around the island visiting familiar spots, we found other alterations, but what didn't show were the political and social changes that had taken place over the years. The whole of Sapelo is now owned by the State of Georgia, except for Hog Hammock, that village area in the center of the island still owned by descendants of slaves manu-

mitted after the Civil War. The entire north end has
been designated the R. J. Reynolds Wildlife Refuge,
and is administered by the Fish and Game Division.
Fifteen hundred acres of the south end has been
leased to the University of Georgia for the Marine
Institute. The rest of the south end is controlled by
the Department of Natural Resources.

The Duplin River is now a federal marine sanc-
tuary and includes the surrounding marshes, Little
Sapelo, and the south beach up to high-tide mark.
You may make arrangements with the Coastal Resources Division of the Georgia Department of Natural Resources in Brunswick for a tour of Sapelo.

The Big House, the old tabby plantation mansion
which was R. J. Reynolds' home when we lived on
the island, is used by the university as a conference
center. Jimmy Carter stayed for a time there during
his presidency. The house remains the same as it
was, including most of the furnishings.

Dramatic changes to both the land and beach occurred with the removal of the large herd of cattle
that used to roam freely over the pasture in the central portion of the south end, through the adjacent
woods, and onto the beach dunes. The cattle were
rounded up and sold on the mainland in the early
1960s. Less than fifty head remain in the interior of
the upper end of the island and they are generally
wary, difficult to see, and don't get to the beach.

The growth of the sand dunes on the south end

was astonishing. When the cattle were taken away, beach grasses could flourish, and the dunes grew up as plants bound the sands captured from the winds. Where they were only inches to a couple of feet high when cattle were present, they now towered above our heads. The beach has also moved seaward so that dunes now stand tall where twenty years ago children paddled in the surf. This resulted from sands eroding from one place and being caught by the grasses in another. The south end sands come from, among other places, the northern part of Cabretta Island, a barrier beach islet about in the middle of Sapelo. This is a natural process. Barrier islands move up and down all along the coast as winds and currents redistribute their sands. As long as people don't interfere too much, the barrier's appearance remains about the same, merely changing its position as winds and currents vary. One changed aspect of Sapelo can be seen from the ocean. The old beach cabanas built by R. J. Reynolds were hidden by the new dunes, and their place has been taken by a large new cabana built by the Department of Natural Resources for Sapelo visitors. But though a large visual change, this is a minor one as far as the battle between beach and sea is concerned. It won't affect the outcome of the continuous adjustments and will eventually become a casualty of them.

Another dramatic change occurred on Sapelo when a large paper plant was opened between Savannah and Brunswick. The plant has a tremendous thirst for water, satisfied by large wells into the aquifer which supplied Sapelo. The day the plant opened and began pumping, the fountain in the laboratory courtyard abruptly stopped flowing.

This occurrence also affected the duck ponds on the north end of the island. There used to be a flow of fresh water from uncapped artesian wells when we lived on the island. We remembered the sulfurous smell from the bubbling, clear water and the growth of slimy but beautifully hued algae around the well top. When this water, which used to flow into and maintain the ponds, was cut off, the ponds had to rely on rain water for refilling. The fish and game managers have solved their problems and even improved the situation, however. They occasionally open the northernmost pond so that salt water can run in. This maneuver keeps down the cattails, preserves some open water, and doesn't seem to affect the animals living there. More herons and ducks inhabit that pond than ever. The second pond, almost cattail-choked twenty years ago, cannot be managed with seawater simply and is now entirely filled with plant growth.

As dramatic as the bird life of Sapelo seemed to us twenty years ago, it seemed so now. In the south end, the loud-mouthed Chacalaca still gives out its rusty call, and the talking among the herons and gulls is still loud. As we drove along a roadway in the center of the island, three wild turkeys flew up from the side of the road directly over the hood of the jeep we were in. The birds seemed to fill our whole field of vision. We found that the wild turkey flocks have increased so greatly that they are caught and exported to other wildlife refuges.

And in a like economic vein, there is a commercially viable operation in the thinning of the planted pine forests which were badly crowded. The cleared fields are cut and kept open to encourage deer, which seemed plentiful. We came upon a buck and two does grazing in the oak park of the south end near the Big House.

The one element that has remained relatively constant has been the marshes. The seasonal laying down of a mat of roots has gone on every season since we left as it had eons before we lived on Sapelo. The high marshes grew at their predictable rate and encroached on the small estuaries. We remembered the hours we had spent looking into the workings of the marsh, the rise and ebb of water over the marsh muds, the busy scrabbling of the fiddler crabs, the schooling of minnows carried in on advancing water. Still standing was the boardwalk over the marsh at the south end. Built of posts driven into the mud with cross pieces nailed on to hold a double row of planks, it seemed about in the same somewhat unstable condition. One of the bonuses of our Sapelo revisited trip was to learn that this somehow still extant structure was named Teal's Boardwalk. It seemed like the sort of immortality you might achieve if you owned a piece of real estate on a Monopoly board.

Sapelo Island will always remain a very special place to us, and if we should be fortunate enough to live another twenty years, we hope it will be possible to visit a third time to again see its changing moods, its land, its beaches, and its marshes.

MILDRED TEAL AND JOHN TEAL
Woods Hole, Massachusetts
February 1981

Index

algae, 99, 106-7, 121, 124, 145-7
alligator, 22, 37, 52, 76-8
amphipod, 131, 143
angel wing, 131
anhinga (water turkey or snake bird), 79, 86-7
annelid, 117, 121, 124, 146; polychaete annelid, 148
anole, 54-5; green, (new world chameleon), 53
ant, 61-2, 111; carpenter, 62; fire, 62, house, 62; pharaoh's ant, 62; red pepper, 62; velvet, 62-3
arachnid, 133
"asbestos buds," 36

baby's ear, 131
baby menhaden, 124
bacteria, 106, 123, 145-6
barn swallow, 113
beach hopper, 144
bee, 63
beetle, 131; dung, 51; June bug, 63; rove, 143; tiger, 143

bird, 75-88, *see* specific entries
bivalve, 149
black skimmer, 154
black widow spider, 59
blood ark, 131
blue crab, 124
blue fish, 124
blue jay, 22-3
brittle star, 151
brown thrasher, 40
bunting, 39-40; painted, 38, 113

cattail marsh, 36-7
chachalaca (Mexican pheasant), 40
chameleon, 24; new world (green anole), 54
channeled duck, 131
chlorophyll, 33
chuck-will's-widow, 43
cicada, 62
clam, 149; stout razor, 131; coquina, 147
cobra, 51
cochina (butterfly shell), 131

171